Cassiane Jrayj de Melo V. Bariani
Nelson M.V. Bariani

Remote Sensing Studies Applied to Commercial Crops

Cassiane Jrayj de Melo V. Bariani
Nelson M.V. Bariani

Remote Sensing Studies Applied to Commercial Crops

Case Studies in Southern Brazil

ScienciaScripts

Imprint
Any brand names and product names mentioned in this book are subject to trademark, brand or patent protection and are trademarks or registered trademarks of their respective holders. The use of brand names, product names, common names, trade names, product descriptions etc. even without a particular marking in this work is in no way to be construed to mean that such names may be regarded as unrestricted in respect of trademark and brand protection legislation and could thus be used by anyone.

Cover image: www.ingimage.com

This book is a translation from the original published under ISBN 978-613-9-72685-1.

Publisher:
Sciencia Scripts
is a trademark of
Dodo Books Indian Ocean Ltd. and OmniScriptum S.R.L publishing group

120 High Road, East Finchley, London, N2 9ED, United Kingdom
Str. Armeneasca 28/1, office 1, Chisinau MD-2012, Republic of Moldova, Europe
Printed at: see last page
ISBN: 978-620-7-91941-3

Summary

PRESENTATION

The book "Studies in Remote Sensing Applied to Commercial Crops" brings together a series of case studies that demonstrate the applications of geotechnologies in the agricultural environment. Currently, the management and monitoring of agriculture is demanding greater control of the production stages, and the use of new technologies in the field is essential, as production costs are rising and producers are working with ever smaller margins.

As such, the producer must seek total control of his activities, monitoring each stage of production, being able to integrate all the variables that involve the production system, seeking to have a vision of the whole while being able to make specific decisions at the right time. One way of achieving these objectives is through geotechnologies, which have enabled farmers to optimise their natural and financial resources, while at the same time making it possible to take decisions in a timely manner and with greater knowledge and availability of information that ranges from the totality of their business to the specifics of an agricultural plot in a given period of time, enabling efficiency that has never been seen before.

Therefore, this book is very timely for the current stage of development and technological advances in agricultural monitoring by orbital remote sensing. It will undoubtedly stimulate the application of these methodologies in many agricultural ecosystems, and will be useful for undergraduates, postgraduates and professionals in agronomy, agricultural engineering, geography, surveying, consultants and decision-makers who need to improve their knowledge and apply it to the management, monitoring and inspection of commercial agricultural areas.

PREFACE

There is almost a consensus that agriculture is one of the sectors undergoing the greatest changes, mainly due to the introduction of geotechnologies. The use of GNSS, precision agriculture, the use of satellites and drones for mapping, monitoring and inspecting crops is an expanding reality.

Agriculture is becoming more and more in the hands of a few large producers. Technological advances are taking place in the sense of control and automation of production processes, requiring increasingly detailed and sophisticated knowledge of what happens on farms. As a result, there is a need to improve techniques that are well-established in academic circles but have only recently been applied in the field. With this in mind, this book presents various case studies and examples of the applicability of remote sensing, geographic information systems and image processing to guide the reader through the practice of monitoring, managing and inspecting agriculture via satellite.

CHAPTER 1

INTERPRETATION OF LANDSAT IMAGES DURING THE RICE DEVELOPMENT CYCLE

Nelson Mario Victoria Bariani, Cassiane
Jrayj de Melo Victoria Bariani.

INTRODUCTION

Agricultural production requires a sequence of actions and synchronisation, from soil preparation to harvest, which need to be carried out at the right time in order to achieve the crop's maximum production potential.

The interpretation of satellite images combined with field information on the agricultural production process has a high potential to help improve the management of irrigated rice plantations.

To analyse this hypothesis, 6 commercial plantations with the IRGA424RI cultivar were monitored by satellite and followed up in the field during the 2016/17 harvest, recording the characteristics of the area, the management procedures and the main stages of rice development throughout the cycle.

Normalised difference vegetation indices (NDVI) from 19 Landsat8/OLI satellite images were processed and analysed during the 2016/17 harvest between September and April. Statistical parameters and photointerpretation of the NDVI images, pixel by pixel (30m spatial resolution), were analysed and discussed with the aid of the panchromatic band (15m resolution) when necessary.

Photointerpretation of the images throughout the cycle made it possible to infer the main characteristics of the area, the plots used in production, the uniformity of the crop, the irrigation channels, the progress of sowing and harvesting. The NDVI values per pixel within each plot could be associated with the phenological stages, and this calibration was tested on a late plot.

It is concluded that the interpretation of NDVI images from the Landsat8/OLI satellite can support the management and monitoring of irrigated rice crops, providing information on crop management and health that can be associated with the truth in the field.

MATERIAL AND METHODS

The study area corresponds to 6 commercial plantations with the IRGA 424RL cultivar Figure 1 shows the areas under study.

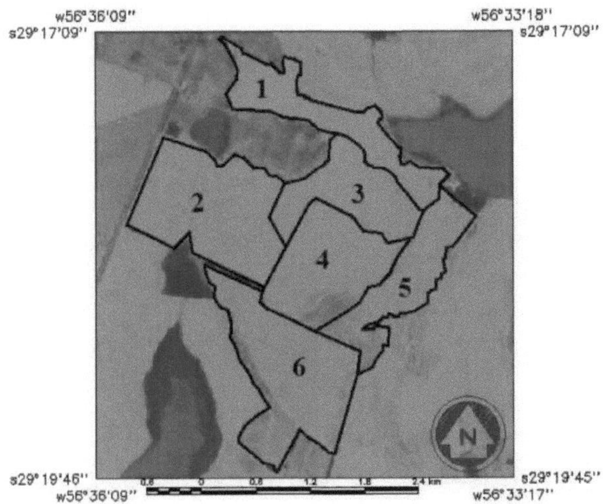

Figure 1 Map of the 6 commercial plantations analysed. Landsat8/OLI band 8 image.

For satellite monitoring throughout the development cycle, 19 scenes were acquired corresponding to orbits 224 and 225 at point 80 of the Landsat8/OLI satellite in the period corresponding to the rice cycle (September to March), in the 2016/2017 harvest, in the municipality of Itaqui, RS.

The images of the area under study were subjected to radiometric and geometric corrections and NDVI extraction (ROUSE *et aL,* 1974). The procedures were carried out using Spring software. Firstly, the registration, contrast and vegetation index extraction procedures were carried out. Secondly, the agricultural plots were vectorised and visually interpreted on screen based on field surveys and photographic records.

5

RESULTS AND DISCUSSION

This research used NDVI extracted from Landsat8/OLI images to interpret the characteristics of the areas, the management used and the stage of development of the IRGA 424RI cultivar in 6 commercial plantations in the municipality of Itaqui, RS, throughout the irrigated rice development cycle. In this way it was possible to visually identify the management of each area, the uniformity of the crop, the irrigation channels, the water intake, the progress of sowing and harvesting. The space-time mapping of NDVI throughout the development cycle of the IRGA 424RI cultivar is shown in Figure 2.

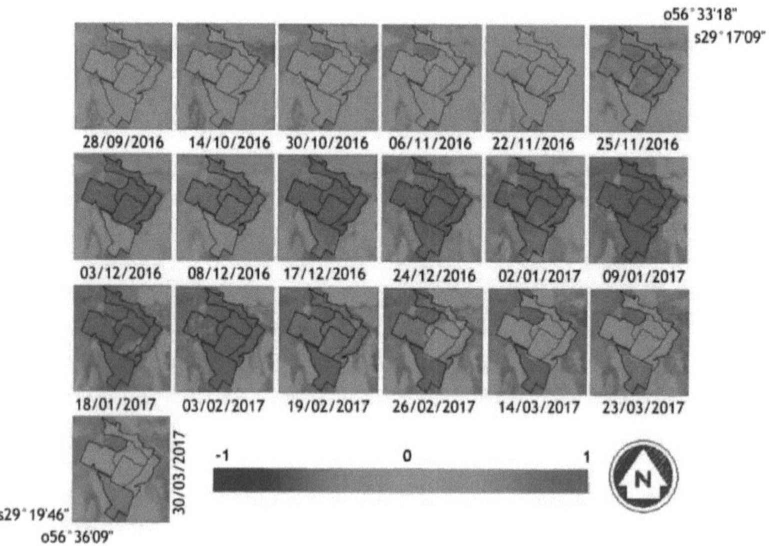

Figure 2 - NDVI images during the cycle of the IRGA 424RI cultivar.

By interpreting the NDVI images (Figure 2), it is possible to assess the development of the IRGA 424RL cultivar

For reasons of better visual discrimination, a chromatic scale of graded tones and three colours, blue, green and red, was used, which was generally associated with the following targets through photointerpretation:

1. Blue indicates targets without vegetation and absorbers of infrared radiation, with

negative NDVI values (-1 to -0.4), such as water from dams or flooded areas;

2. Green indicates NDVI values around zero (-0.4 to 0.4), corresponding to targets with infrared reflectance similar to red, sometimes higher and sometimes lower, such as exposed soils or soils with little vegetation cover or shallow water mirrors or clouds and atmospheric aerosols;

3. Red indicates intermediate to high NDVI values, corresponding to vegetation taller than grass, covering most of the ground, but developed to different degrees, indicated by the intensity of the colour, with the tallest and densest vegetation corresponding to the most intense red.

In general, Figure 2 shows the progressive increase in vegetation cover over the course of the cycle, with a subsequent decrease in NDVI due to senescence and the advance of the harvest.

The first 4 images of the cycle include soil preparation, sowing and the start of vegetative development characterised by green colouring, due to the low NDVI values in this period, as also described by Nobre, (2010). On 22/11/2016, it is already possible to see the start of emergence, marked by the red colour in the lower corner of crops 2 and 4. Complete emergence (reddish colour) occurs on 03/12/2016, where only plot number 6 is not emerging (greenish colour), because it was being sown. On 08/12/2016, water enters the field and the red colour darkens, indicating a decrease in NDVI values for plots 1, 2, 3, 4 and 5. At this stage, the first nitrogen application must have already taken place before the water came in (SOSBAI, 2016).

The images from 17/12/16 to 03/02/17 mark the reproductive stage in crops 1,2,3,4 and 5 (deep red colour) and high NDVI values, also evidenced by Hargrove, *etaL,* (2010) and Wang, *etaL,* (2015). Crop number 6 began its reproductive period on 09/01/17 and lasted until 14/03/17.

In the image from 19/02/17, the cultivar is entering senescence in plots 1,2,3,4 and 5, and is ready to be harvested from this date onwards, all that is needed is the ideal humidity, close to 22% for the grain to be harvested (SOSBAI, 2016). Harvesting in these plots began on 26/02/17, when NDVI values returned to lower levels, as evidenced by the green colouring in part of the plots. On 14/03/17 plots 1,

2, 3, 4 and 5 were already harvested (completely green).

While in the images from 23/03/17 and 30/03/17, crops number 1, 2, 3, 4 and 5 had already been harvested (green), crop number 6 was entering senescence (reddish colour), where it is possible to compare the same behaviour that occurred in the other crops on 19/02/17.

From the above, there is evidence that photointerpretation of NDVI images from the Landsat8/OLI satellite over the development period of the IRGA 424RI cultivar can be a tool to support crop monitoring and management.

CONCLUSION

There is evidence that the visual interpretation of NDVI images throughout the irrigated rice cycle can support the monitoring and management of commercial crops. In order to use images from sensors on board satellites for agricultural monitoring purposes, it is necessary to obtain as many cloud-free images as possible during the harvest. In this work, the fact that the study area overlapped two Landsatô satellite orbits made it easier to take scenes, totalling nineteen images for monitoring the harvest.

As such, Landsat8/OLI satellite images, when in areas of overlapping orbits, become a suitable tool for agricultural monitoring and management.

BIBLIOGRAPHICAL REFERENCES

ALLEN, R. G.; PEREIRA, L. S. Estimating crop coefficients from fraction of ground cover and height. **Irrig. Sei.,** 28, 2009. 17-34.

HARGROVE, W. W. et al. Toward a National Early Warning System for Forest Disturbances Using Remotely Sensed Land Surface Phenology. **USGS,** 2010. Available at: <https://www.geobabble.org/~hnw/first/ncdc/slideshow.html>. Accessed on: 13 June 2017.

NOBRE, F. L. D. L. **Spectrotemporal characterisation of irrigated rice crops using MODIS images.** PELOTAS: UFPEL, 2010.

ROUSE, J. W. et al. **Monitoring Vegetation Systems in the Great Plains with ERTS.** Third Earth Resources Technology Satellite-1 Symposium. Greenbelt: NASA.

1974.

SOSBAI, R. T. D. C. D. A. I. **Irrigated rice:** technical research recommendations for southern Brazil. ISBN 978-85-69582-02-1. ed. Pelotas: [s.n.], 2016. 200 p.

WANG, J. et al. Estimation of rice phenology date using integrated HJ-1 CCD and Landsat-8 OLI vegetation indices time-series images. **Journal of Zhejiang University-SCIENCE B,** v. 16, 14 October2015. p. 832-844.

CHAPTER 2

MONITORING THE PHENOLOGICAL STAGES OF SOYA BEANS USING REMOTE SENSING

Cassiane Jrayj de Melo Victoria Bariani,

Nelson Mario Victoria Bariani,

INTRODUCTION

Among the main factors contributing to the increase in cultivated area and soya productivity in RS are the genetic improvement of cultivars, soil, crop and irrigation water management and, lately, advances in crop monitoring techniques. The normalised difference vegetation index (NDVI), obtained using remote sensing techniques, can be used to provide up-to-date and historical information on a given crop, such as density and vigour, without the need to visit the field, without interfering directly or indirectly in the growth and development of the plants and reducing the costs of determination.

With NDVI it is possible to estimate the phenological stage of the soya crop, as this index can be directly correlated with the amount of vegetation biomass, leaf area index (LAI) and, consequently, with crop transpiration. It provides information on medium and long-term water stress, as it is sensitive to phenomena that take a long time to appear, such as leaf yellowing (PÔÇAS, PAÇO, *et aL,* 2015). In this way, NDVI can help calculate irrigation and estimate crop coefficients, because once you know the phenological stages of the crops, you can calculate the soil cover fraction (Fc), the density coefficient (Kd), the basal crop coefficient (Kcb) and the crop coefficient (Kc). These coefficients, when multiplied by the reference evapotranspiration (ETo), result in the crop evapotranspiration (ETc), where ETc = Kc*ETo, necessary for estimating crop water requirements.

The aim of this study is to differentiate the main phenological stages of soya crops using NDVI in fields irrigated by central pivot in the municipality of Cruz Alta in

Rio Grande do Sul. To achieve this objective, temporal NDVI profiles were generated during the soya development cycle in the 2004/2005 harvest, in 11 irrigation pivots, using 6 images from the LANDSAT 5/TM satellite, in orbits 222/80 and 223/80.

MATERIALS AND METHODS

The area of interest in this study is located in the north-western region of the state of Rio Grande do Sul, in the municipality of Cruz Alta. The region under study is characterised by its extensive agricultural areas, where centre pivot irrigation is one of the region's economic strategies. On the map in Figure 1, the 11 pivots analysed in this study are highlighted in red.

Figura 1. Location of the study area.

Images from the TM *(Thematic Mapper)* sensor on board the Landsat 5 satellite are used in this study, covering the periods from November 2004 to March 2005, totalling 6 images in orbits 222 and 223, at point 80, covering the entire soya crop cycle, from sowing to harvest. The 2004/2005 crop year was chosen due to the greater frequency of imaging, where all the months of the cycle were covered by images, with the month of December having two images (Table 1). Working with two orbits is justified by the fact that it improves temporal resolution, i.e. the frequency with which the satellite revisits the study area. As can be seen in Table 1, there are three images from orbit 222 and three images from orbit 223, thus achieving

11

complete imaging of the soya cycle, doubling the number of images compared to using a single orbit, which is suitable for agricultural monitoring and detecting the phenological stages of crops.

Table 1. Dates of the images with their respective orbits

Date of images	Orbit
23/11/2004	223
02/12/2004	222
18/12/2004	222
26/01/2005	223
27/02/2005	223
08/03/2005	222

The images were digitally processed using Spring software (version 5.2.6) (CAMARA, 1996). The images were pre-processed to check and harmonise them in terms of geo-referencing, projection, the occurrence of clouds and faults, pixel alignment and reflectance in accordance with Markham and Barker (1987a). The Landsat 5 images were obtained free of charge via the internet from INPE's <http://www.dgi.inpe.br/CDSR/> website. Images from the Geocover Mosaic (USGS, 2004) were used to georeference the images.

After the Landsat satellite images have been corrected, the NDVI must be calculated (ROUSE, HAAS, *et al.*, 1974).

$$NDVI = \frac{\rho_{nir} - \rho_{red}}{\rho_{nir} + \rho_{red}} \tag{1}$$

Where ρ_{nir} is the reflectance value in the near infrared range; and ρ_{red} is the reflectance value in the visible red range.

RESULTS AND DISCUSSION

NDVI was processed and extracted from six Landsat 5 satellite images in 11 pivots throughout the soya crop cycle. The pivot identified as number nine was monitored in the field in terms of sowing, harvesting, phenological stage and plant height.

There was a constant increase in the average NDVI until the R5 stage, at 64 days after sowing (DAS), when the maximum point was reached, and it began to decrease from this point onwards. The average NDVI value observed and identified as the maximum point was 0.85, which represented an increase of 66.82% compared to the VE-VC stage (0.17) (Table 2). A soya plant in the VE phenological development stage is characterised by having cotyledons above the soil surface, in the VC stage it has unifoliolate leaves with the margins no longer touching, while in R5 it is in the grain filling stage, with 3 mm grains in a legume from the last four stem nodes with developed leaves (THOMAS and COSTA, 2010).

The increase in average NDVI values as the plant develops is directly related to the increase in its biomass, leaf area index and fraction of ground cover by the vegetation (ALLEN and PEREIRA, 2009).

Table 2. Results of the NDVI values in each centre pivot for each date of the TM sensor images analysed, with their respective days after sowing (DAS) and crop development stages (FEHR, NEUMAIR and NEPOMUCENO, 1977).

Pivot	Landsat 5/TM images					
	23-Nov-04	2-Dec-04	18-Dec-04	26-Jan-05	27-Feb-05	8-Mar-05
5	0.16	0.23	0.36	0.85	0.83	0.79
9	0.18	0.19	0.25	0.84	0.71	0.37
11	0.17	0.23	0.36	0.84	0.82	0.78
12	0.16	0.17	0.32	0.85	0.78	0.78
16	0.16	0.22	0.41	0.85	0.84	0.79
17	0.19	0.22	0.30	0.83	0.83	0.81
20	0.16	0.18	0.31	0.84	0.83	0.80
23	0.18	0.17	0.24	0.83	0.83	0.81
24	0.19	0.20	0.25	0.83	0.82	0.80
26	0.15	0.17	0.30	0.85	0.84	0.80
28	0.21	0.31	0.44	0.83	0.54	0.35
DAS	0	**9**	25	64	97	**106**
EF	**VE-VC**	V2-V3	V4-V5	R1-R5	R6-R7	R8

NDVI values ranged from 0.15-0.21 for the sowing, germination and emergence period; 0.17-0.31 at 9 DAS, corresponding to vegetative stages V2-V3; 0.24-0.44 at 25 DAS, corresponding to stages V4-V5; 0.83-0.85 at 64 DAS, corresponding to stages R1-R5; 0.54-0.84 at 97 DAS, for stages R6-R7 and 0.35-0.81 at 106 DAS, corresponding to stage R8 (physiological maturity). The graph below (Figure 2) shows the temporal variation in NDVI for the soybean cycle in the 2004/2005 harvest for the 11 pivots analysed. For pivot number nine, we have data from surface observations during the crop development cycle, in order to corroborate the data collected via satellite.

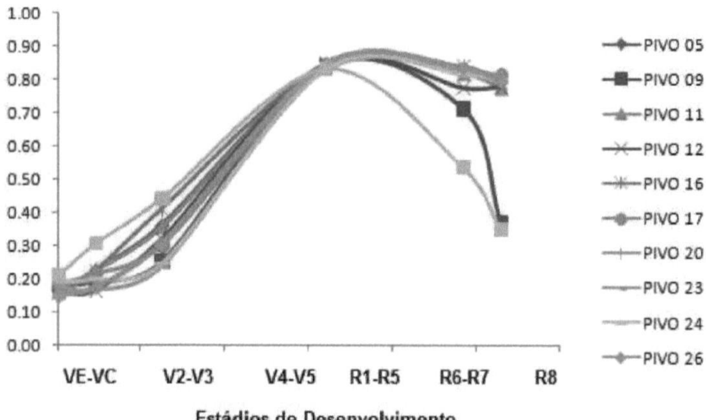

Figura 2. Graph of the NDVI time profile for the soybean cycle in the 2004/2005 harvest for the 11 pivots analysed.

The reduction in average NDVI values at stage R6 (97 DAS) was due to the yellowing of the leaves and a reduction in the photosynthetically active area due to the onset of leaf senescence. The R6 stage is characterised as the final stage of grain filling, where the maximum volume of grains occurs, the pods still have at least one green grain that occupies the entire cavity, in one of the last four nodes of the stem with developed leaves (THOMAS and COSTA, 2010).

From stage R5 (64 DAS) to R8 (106 DAS), there was a 12.44 per cent reduction in the NDVI value. At stage R7, the plants show characteristics of early ripeness, mainly the yellowing of the pods. The start of the stage is considered to be when at least one pod on the stem has reached ripe colour (THOMAS and COSTA, 2010).Applying a trend line to the average NDVI values throughout the soya cycle (Figure 3), a relationship is observed between the NDVI data and soya development, where NDVI is associated with the plant's stage of development. The trend line tends to rise in the vegetative phase and the beginning of the reproductive phase, decreasing towards the end of the cycle, when there is a reduction in the plant's photosynthetic active area (NEIVERTH, CRUSIOL, *et aL,* 2013).

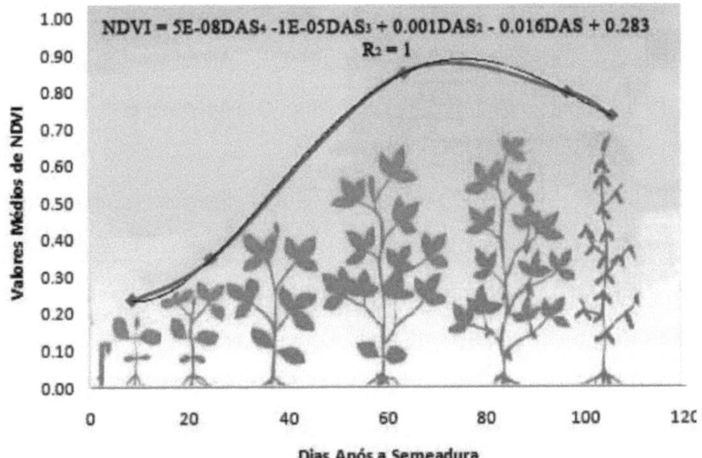

Graph of the average values for the NDVI time profile for the soybean cycle and their respective phenological stages for the 2004/2005 harvest in 11 analysed centre pivots.

The maximum NDVI point observed (0.85) occurred at 64 DAS, with soya between stages R1-R5; studies carried out in a greenhouse also show a maximum NDVI point (0.90) at 69 DAS with soya at stage R5.4 (NEIVERTH, CRUSIOL, *etaL,* 2013). There is a drop in NDVI values at stages R6 and R7 due to crop maturation and at stage R8 due to harvesting, which for some pivots occurred at 106 DAS with an average NDVI value of 0.72.

CONCLUSION

The use of NDVI allowed the main phenological stages of the soya crop to be differentiated in fields irrigated by centre pivot in the municipality of Cruz Alta in Rio Grande do Sul. Further studies are recommended, extending the NDVI time series to other crop years, to effectively support the planning and monitoring of the soya crop in terms of phenological stages, to effectively replace field monitoring and support irrigation management.

BIBLIOGRAPHICAL REFERENCES:

ALLEN, R. G.; PEREIRA, L. S. Estimating crop coefficients from fraction of ground cover and height. **Irrig. Sei.,** 2009. 17-34.

ANTUNES, J. F. G.; ESQUERDO, J. C. D. M. **Agricultural monitoring using harmonic analysis of NDVI/AVHRR-NOAA data time series.** XIV Brazilian Symposium on Remote Sensing. Natal: INPE. 2009. CÂMARA, G. Intergrating remote sensing and GIS by object-oriented data modelling. **Computers & Graphics,** v. 20, n. 3, p. 395-403, May 1996.

FEHR, W. R. B.; NEUMAIR, N.; NEPOMUCENO, A. L. **Stages of soybean development.** [S.l.]: Ames: Lowa State University, 1977.

MARKHAM, B. L.; BARKER, J. L. Radiometric Properties of U.S. processes Landsat MSS data. **Remote Sensing of Environment,** New York, 17, 1987a. 39-71.

MARKHAM, B. L.; BARKER, J. L. Thematic Mapper bandpass solar exoatmospherical radiances. **International Journal or Remote Sensing,** 8, n. 3, 1987b. 517-523.

NEIVERTH, W. et al. **NDVI of phenological stages of soybean BRS 284.** VIII Embrapa Soja Academic Conference. Londrina, PR: [s.n.]. 2013. p. 204-209. PÔÇAS, I. etal. Estimation of Actual Crop Coefficients Using Remotely Sensed Vegetation indices and Soil Water Balance Modelled Data. **Remote Sensing,** 2015. 1-29.

ROUSE, J. W. et al. **Monitoring Vegetation Systems in the Great Plains with ERTS.** Third Earth Resources Technology Satellite-1 Symposium. Greenbelt: NASA. 1974.

THOMAS, A. L.; COSTA, J. A. **Soya:** Management for high grain productivity. Porto Alegre: Evangraf, 2010.

USGS. GLCF: Landsat GeoCover Mosaic. **Global Land Cover Facility,** 2004. Available at: <http://glcf.umd.edu/data/mosaic/>. Accessed on: 29 September 2014.

CHAPTER 3

REMOTE SENSING TO STUDY VEGETATION DYNAMICS IN IRRIGATED AREAS

Cassiane Jrayj de Melo Victoria Bariani,
Nelson Mario Victoria Bariani.

INTRODUCTION

The use of remote sensors to map vegetation cover, land use and the phenological stage of crops is the subject of growing interest from public and research organisations. Even in countries with the largest water reserves in the world, such as Brazil, entrepreneurs and managers maintain the policy and effective economic practice of using natural resources to their limits. From this point of view, forestry and water resource agencies, as well as water users, have frequent demands for mapping at a resolution capable of contributing to crop management, monitoring the health of forestry and water resources, assessing water consumption, among other applications.

Satellite data is valuable for evaluating vegetation, soil and water information. A common strategy for using this information is to transform it into vegetation indices used to highlight the differences between soil and vegetation. One of the most widely used vegetation indices is NDVI, which is a sensitive indicator of the quantity and condition of vegetation. This index synthesises the non-redundant information contained in the different images (bands) obtained by the satellite sensor. This is because the detection of targets corresponds to different parts of the electromagnetic spectrum, in which the responses of each terrestrial object vary in magnitude and shape, containing information specific to their physical and chemical characteristics. Because of this purpose, its calculation benefits, in terms of quality and precision, from the conversion of the raw values contained in the pixels (image elements) to physical values that emphasise the structural information (composition, texture) of the targets, and not geometric characteristics such as inclination in relation to the sun. In other words, it is beneficial for the vegetation index values within an image to show differences due to the composition and internal characteristics of the targets (vegetation and soils, for example), and not differences due to the angle of reflection of the incident solar radiation.

The processing of satellite images, including their registration (georeferencing), colour compositions, vector editing, matrix operations and index calculation can be carried out in various Geographic Information System software packages available today, one of which is the Georeferenced Information Processing System (SPRING), used in this work. The first stage of image processing is georeferencing, followed by radiometric transformation, i.e. converting the Digital Number (ND) to apparent reflectance values (at the top of the atmosphere). This strategy is adopted to reduce the variability of inter-cene responses, correct the images for the effect of differences in the solar zenith angle (due to the different times the images were taken) and compensate for differences in solar irradiance at the top of the atmosphere (CHANDER, MARKHAM and BARSI, 2007).

The aim of this study is to evaluate and discuss the entire calculation routine that converts raw data (digital numbers) into physical values (radiance and reflectance) in centre-pivot irrigated areas. In this way, the aim is to obtain NDVI values that are more consistent with reality, as well as emphasising the importance of converting raw values into physical values before determining vegetation indices so as not to compromise the final NDVI result, as well as other calculations and procedures that use this information, such as identifying the phenological stages of crops.

MATERIALS AND METHODS

The area of interest in this study is located in the north-western region of the state of Rio Grande do Sul, in the municipality of Cruz Alta (Figure 1). The region has the largest area irrigated by centre pivot in RS. The geology is predominantly basic igneous rocks, mainly basalt, with the rest of the soil developed from Botucatu sandstone or *a* mixture of this and basalt (STRECK, KAMPF, *et a!,* 2008).

Figure 1 - Location of the study area.

Remote Sensing Products and Data

Images from the TM *(Thematic Mapper)* sensor on board the Landsatõ satellite were used in this study, covering the periods from January to December 2004, totalling 17 images in orbits 222 and 223, at point 80. The year 2004 was chosen due to the greater frequency of imaging, where practically every month of the year was covered by images, and in some months there was more than one image. Working with two orbits is justified by the fact that it improves temporal resolution, i.e. the frequency of satellite revisits to the study area, which went from 16 days to 8 days. As can be seen in Figure 1, the orbits have an overlap of 15 km for the 80th point analysed, obtaining a complete annual image that enables agricultural monitoring and the detection of crop phenological stages.

The reason for using the Landsatõ satellite, apart from the fact that the images are available free of charge, is that it has a frequency for updating its calibration data, as well as being concerned with publicising it. This allows users to establish correlations between the radiometric parameters determined by applying this calibration data (CHANDER, MARKHAM and BARSI, 2007).

The images were digitally processed using Spring software version 5.2.6

(CAMARA, 1996). The images were pre-processed to check and harmonise them in terms of georeferencing, projection, the occurrence of clouds and faults and reflectance. The Landsat 5 images were obtained free of charge via the internet from the INPE website <http://www.dgi.inpe.br/CDSR/>. Images from the Geocover Mosaic (USGS, 2004) were used to georeference the images.

Collection of conversion equations

According to Markham and Barker (1987a) the digital number (ND) must be converted into radiance according to the equation:

$$L_\lambda = \left(\frac{L_{máx}-L_{mín}}{ND_{máx}-ND_{mín}}\right).(ND - ND_{mín}) + L_{mín} \tag{4}$$

Where, ND is the digital number in each pixel; $L_{máx}$ and $L_{mín}$ are calibration constants for a given sensor; $ND_{máx}$ and $ND_{mín}$ are the maximum and minimum values that ND can reach; L_λ is the monochromatic spectral radiance (W/m^2 .sr.pm).

The TM sensor on the Landsatõ satellite has $ND_{mín}$ = 0 and $ND_{máx}$ = 255, as its radiometric resolution is 8 bits, i.e. 256 grey levels, which corresponds to a scale of 0 to 255 grey levels. Radiance can therefore be calculated using the following formula:

$$L_\lambda = \left(\frac{L_{máx}-L_{mín}}{ND_{máx}}\right).(ND) + L_{mín} \tag{5}$$

And the monochromatic reflectance can be found by the equation:

$$\rho_\lambda = \frac{\pi.L_\lambda}{E_\lambda.\cos(z).d_r} \tag{6}$$

Where, d_r is the inverse of the square of the Earth-Sun relative distance in astronomical units; z is the zenith sun angle (degrees) at the time of acquisition; E_λ is the average solar irradiance at the top of the atmosphere (mW/cm^2 .Ω.µm); L_λ is the monochromatic spectral radiance (W/m^2 .sr.µm); and ρλ is the monochromatic reflectance.

To obtain the conversion directly from ND to reflectance, you just need to turn equation 5 into 6, as shown below:

$$\rho_\lambda = \frac{\left[\left(\frac{L_{máx}-L_{mín}}{ND_{máx}}\right).(ND)+L_{mín}\right]}{E_\lambda.\cos(z).d_r} \tag{7}$$

After the Landsat satellite images have been converted to physical values, the NDVI must be calculated (ROUSE, HAAS, *etal.*, 1974).

$$NDVI = \frac{\rho_{nir}-\rho_{red}}{\rho_{nir}+\rho_{red}} \tag{8}$$

Where ρ_{nir} is the reflectance value in the near infrared range; and ρ_{red} is the reflectance value in the visible red range.

Data collection for the conversion

The data that will be applied in the calculation routine described above will be explained below:

The solar zenith angle (β) (equation (7)) is determined by the difference between the zenith and the elevation of the Sun (β), data extracted from the image header. For the scenes used we have: Sun elevation (/3) in decimal degrees (Table 1). Thus the cosine of the solar zenith angle is determined by the equation below:

$$\cos (z) = \cos (90-\beta) \tag{9}$$

Table 1. Dates of the TM sensor images with their corresponding Sun Elevation data and the result of equation (9).

Date of TM/L5 images	Elevation of the Sun (β)	Result Eq. (9)
01/01/2004	56.3229	0.832176
18/02/2004	49.3761	0.759
06/04/2004	40.0007	0.642797
13/04/2004	38.4416	0.621717
09/06/2004	27.9558	0.46879
11/07/2004	27.9227	0.46828
27/07/2004	30.1635	0.502469
03/08/2004	31.5859	0.523776
12/08/2004	33.7795	0.555998
04/09/2004	40.7874	0.653254
29/09/2004	49.3276	0.758448
06/10/2004	51.6052	0.78375
15/10/2004	54.2748	0.811827

07/11/2004	57.777	0.845979
23/11/2004	60.4538	0.869958
02/12/2004	60.4282	0.869738
18/12/2004	59.3776	0.860543

According to (BASTIAANSSEN, 1995) the term d_r is defined as $1/d^2_{e\text{-}s}$ where $d_{e\text{-}s}$ is the relative Earth-Sun distance in astronomical units (ALLEN, PEREIRA, *et aL,* 1998) states that d_r is obtained from the following equation:

$$d_r = 1 + 0.033 \cos\left(DJ \frac{2\pi}{365}\right) \tag{10}$$

Where DJ represents the Julian Day (sequential day of the year), for the scenes under study, the dates of the images and the corresponding Julian Days are shown in Table 2, along with the result of equation (10).

Table 2. Dates of Landsat/TM images with their corresponding Julian Day and result of equation (10).

Image date	Julian Day	Result Eq. (10)
01/01/2004	1	1.032995
18/02/2004	49	1.021997
06/04/2004	97	0.996889
13/04/2004	104	0.992973
09/06/2004	161	0.969326
11/07/2004	193	0.967485
27/07/2004	209	0.970233
03/08/2004	216	0.972155
12/08/2004	225	0.975213
04/09/2004	248	0.9855
29/09/2004	273	0.99915
06/10/2004	280	1.003111
15/10/2004	289	1.00813
07/11/2004	312	1.019807
23/11/2004	328	1.026224
02/12/2004	337	1.028994
18/12/2004	353	1.032182

Table 3 (below) describes data from bands 3 and 4 of Landsat 5 - TM, used to calculate NDVI. The calibration coefficients *(min L and max L)* were taken from (CHANDER, MARKHAM and BARSI, 2007). The spectral irradiances at the top of the atmosphere were taken from Markham and Barker (1987b).

Table 3. Description of Calibration Coefficients and spectral irradiance at the top of the atmosphere.

Bands	Calibration coefficient		Solar irradiance at the top of the
	Lmin	Lmax	atmosphere
3	-1.17	264	1554

| 4 | -1.51 | 221 | 1036 |

Source: Adapted from Chanderef *al.*, (2007) and Markham and Barker (1987b).

Spatial Language for Algebraic Geoprocessing (LEGAL)

Proposed by (CÂMARA, 1996), the Spatial Language for Algebraic Geoprocessing (LEGAL) is part of the SPRING package developed by INPE. It consists of a list of sentences that describe a procedure, i.e. a set of actions on spatial data, that makes sense in the context of some discipline of Geographic Information Systems (NETO, BARROS FILHO, *et al.*, 2008). This work seeks to use LEGAL programming resources, part of the SPRING software in version 5.2.6, to determine the reflectance values of each pixel in the image under study and convert them to real physical values.

RESULTS AND DISCUSSION

The algorithm described below, which follows a LEGAL routine, was devised to convert ND into reflectance. This algorithm can be used for any LANDSAT5/TM image, simply by obtaining the solar zenith angle and Julian day data for each image. The algorithm in question was developed using data from the image of 29 September 2004 from orbit 222 at point 80.

```
{
//CÁLCULO REFLETÂNCIA e NDVI LANDSAT5
//Imagem: 2004-09-29 L5 TM 222/80
// Declaração de variáveis (tipo, nome e modelo)
Imagem B3, B4 ("Landsat");
Numerico refletancia3, refletancia4 ("Reflectancia");
// Instanciação (definição) de variáveis reais
pi = 3.1415926;
cosz = 0.758448;
d = 0.99915;
//Parâmetros de calibração do LANDSAT 5Markham (2009)
lmin3 = -1.17;
lmax3 = 264;
lmin4 = -1.51;
lmax4 = 221;
e3 = 1536;
e4 = 1031;
```

```
//Números digitais máximos: 255 no LANDSAT5
ndmax = 255;
// Instanciação de variáveis
B3 = Recupere (Nome="B3");
B4 = Recupere (Nome="B4");
refletancia3 = Novo (Nome = "29092004refl_B3",ResX = 30,ResY = 30,Escala = 100000,Min = 0, Max = 50);
refletancia4 = Novo (Nome = "29092004refl_B4", ResX = 30,ResY = 30, Escala = 100000,Min = 0, Max = 50);
//Por fim as operações desejadas
// Conversão de ND para reflectância
refletancia3 = Numerico ((pi/(d*e3*cosz))*((B3/ndmax)* (Imax3-Imin3)+ Imin3));
refletancia4 = Numerico ((pi/(d*e4*cosz))*((B4/ndmax)* (Imax4-Imin4)+ Imin4));
//Elaboração do Mapa Temático
Numerico NDVI("Grade_NDVI");
TematicoVegRefl("Vegetacao");
NDVI = Novo (Nome = "29092004_NDVI", ResX = 30, ResY= 30, Escala = 100000, Min = -1, Max = 1);
NDVI = ((refletancia4 - refletancia3)/( refletancia4 + refletancia3));
Tabela fatias(Fatiamento);
fatias = Novo (CategoriaFim = "Vegetacao",
[-1.0,0.0] : "Sem Vegetação",
[0.0,0.2] : "Estádio Vegetativo/Senescência",
[0.2,0.5] : "Estádio Reprodutivo",
[0.5,1] : "Fase de Colheita");
VegRefl = Novo (Nome="EF29092004",ResX=30,ResY=30,Escala=100000);
VegRefl = Fatie (NDVI, fatias);
}
```

After running the above programme, reflectance data was generated and transformed into an image and grid with NDVI values (Figure 3a). Based on the NDVI information, thematic maps of crop phenology were drawn up for each date imaged during the period studied (Figure 3b).

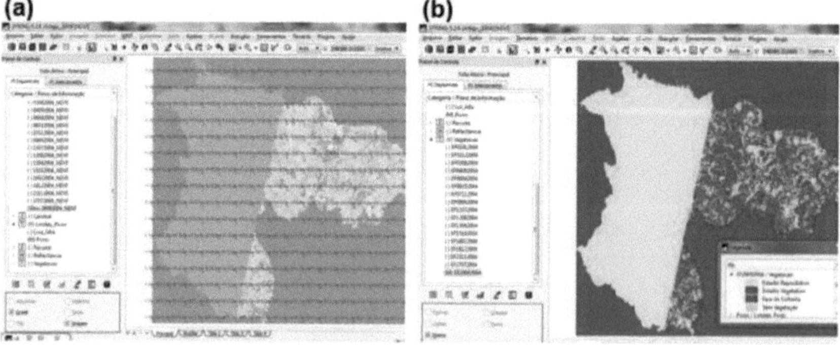

Figura 3. Processing result - algorithm in LEGAL, (a) Image and grid of NDVI values resulting from the execution of the first part of the algorithm presented; (b) Thematic image of NDVI values resulting from the slicing proposed by the execution of the second part of the algorithm presented.

Figure 5 shows the thematic maps corresponding to the 17 images taken in 2004. The NDVI values have been sliced, i.e. grouped so that they are closer to the crop stages. In pink are the values between -1 and 0 corresponding to bare soil; in red are the values between 0 and 0.2 corresponding to the vegetative stage; in yellow are the values between 0.2 and 0.5 corresponding to the reproductive stage of the crop; and in green are the values between 0.5 and 1 corresponding to the harvest stage.

For a better discussion of the results, information was obtained on the sowing and harvesting of some crops in some pivots in the study area, as shown in Table 4.

Table 4 - Type of crop, sowing date and crop closure date - information obtained from the producer's history.

Pivot	Culture	Sowing date	Harvest date
Pivot 09	Soya	22/11/2004	07/03/2005
Pivot 12	Maize	04/09/2003	27/01/2004
Pivo 27	Maize	01/10/2004	15/02/2005

The information from the field monitoring showed that the pivots with the sequence green in January and yellow in February represent the maize crop that was harvested at the end of January, according to the field monitoring (27/01/2004); while the pivots with a sequence of green in January and green in February, or yellow in January and green in February, are pivots with soya.

Through these observations it was possible to identify the centre pivots that grew soya and maize. The centre pivots that grow soya correspond to 13 pivots: 4, 5, 8, 9, 13, 18, 21, 23, 24, 25, 27, 28, 30. The pivots that grow maize are: 1, 2, 3, 6, 7, 10, 11, 12, 14, 15, 16, 17, 19, 20, 22, 26 and 29.

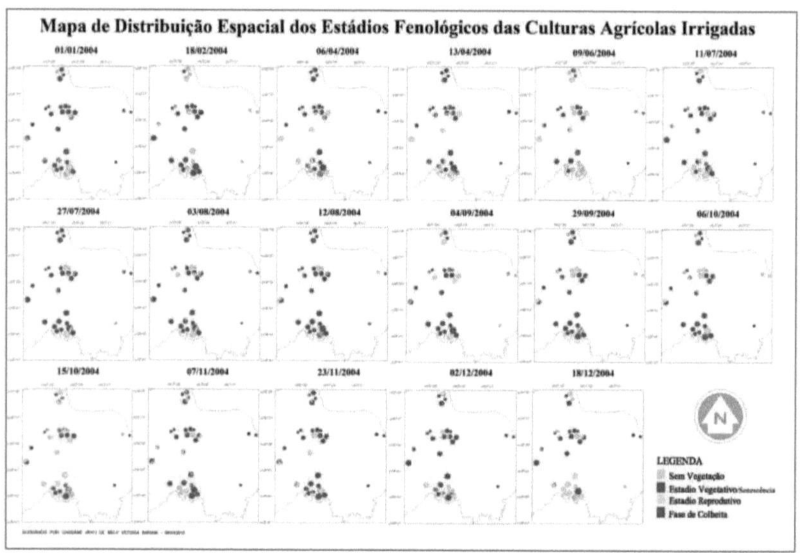

Figure 5 - Map of the distribution of NDVI values for the 17 images analysed in the 30 central irrigation pivots.

CONCLUSION

The methodology presented made it possible to convert the raw data (digital numbers) into physical values (radiance and reflectance), consistent with the reality of the 30 centre pivots. The NDVI generated from the reflectances made it possible to identify the main stages of crop development. There is evidence that NDVI values between 0.0 and 0.2 represent the vegetative or senescence stage of the crop, between 0.2 and 0.5 the reproductive stage and above 0.5 the harvest or maturity stage. This data can be verified through the historical records of producers in 3 pivots, which made it possible to identify 17 pivots with corn crops planted and 13 with soya crops.

BIBLIOGRAPHICAL REFERENCES

ALLEN, R. G. et al. FAO Irrigation and drainage paper No. 56. In: FAO **Food and Agriculture Organisation of the United Nations.** Rome: [s.n.], 1998. p. 26-40.

BASTIAANSSEN, W. G. M. **Regionalisation of surface flux densities and.** [S.l.]: Ph.D. Thesis, 1995.

CÂMARA, G. Intergrating remote sensing and GIS by object-oriented data modelling. **Computers & Graphics,** v. 20, n. 3, p. 395-403, May 1996.

CHANDER, G.; MARKHAM, B. L.; BARSI, J. A. Revised Landsat-5 Thematic Mapper Radiometric Calibration. **IEEE Geoscience And Remote Sensing Letters,** 4, n. 3, 2007. 490-494.

NETO, R. T. D. B. et al. **DETERMINATION OF PHYSICAL VALUES OF TM/LANDSAT-5 IMAGES USING LEGAL LANGUAGE TO OBTAIN VEGETATION INDEXES.** II Brazilian Symposium on Geodetic Sciences and Geoinformation Technologies. Recife: [s.n.]. 2008. p. 1-8.

ROUSE, J. W. et al. **Monitoring Vegetation Systems in the Great Plains with ERTS.** Third Earth Resources Technology Satellite-1 Symposium. Greenbelt: NASA. 1974.

STRECK, E. V. et al. **Soils of Rio Grande do Sul.** 2. ed. Porto Alegre: EMATER/RS, v. 1, 2008. 222 p.

USGS. GLCF: Landsat GeoCover Mosaic. **Global Land Cover Facility,** 2004. Available at: <http://glcf.umd.edu/data/mosaic/>. Accessed on: 29 September 2014.

CHAPTER 4

SPATIAL VARIABILITY DURING THE DEVELOPMENT CYCLE OF IRRIGATED RICE

Cassiane Jrayj de Melo Victoria Bariani,
Nelson Mario Victoria Bariani.

INTRODUCTION

Rice (Oryza sativa L.) is one of the most important crops, being the second most cultivated cereal in the world, and the first in direct consumption by humans. More than 1 billion people depend on rice as their staple food and means of subsistence. However, rice production faces major challenges over the next decade due to rising production costs, natural resource pressures, climate change, population growth, changing eating habits and economic development (TORBICK, *et al.*, 2017).

There is a strong need for an effective means of monitoring rice cultivation, providing and improving information on the crop and growing conditions to support decision making in advance of possible falls in productivity that may occur.

Remote sensing provides the essential technology and methodology for monitoring, mapping and observing rice crops, and is capable of providing timely and reliable information for various purposes related to rice farming systems (KUENZER and KNAUER, 2013), at sufficient time intervals to interpret changes that could cause economic damage and a drop in productivity.

This research suggests analysing the variability of the normalised difference vegetation index (NDVI) in irrigated rice fields throughout the crop's development cycle in order to support strategic planning and decision-making in agricultural enterprises, as it can serve as an indicator of the need for management.

MATERIAL AND METHODS

This study monitored three irrigated rice fields with the cultivars IRGA 429, in a field covering 24 hectares, and IRGA 424RI in two fields, one covering 60 and the other 100 hectares (Figure 1), in the municipality of Itaqui, located on the western border of the state of Rio Grande do Sul, Brazil.

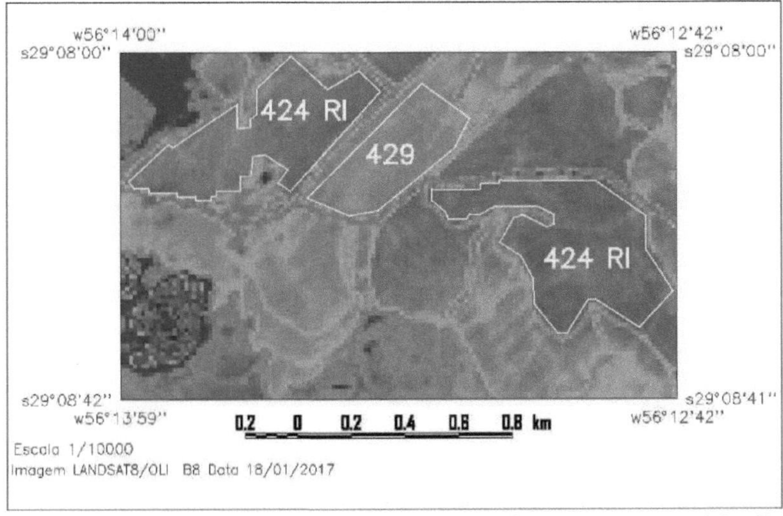

Figura 1 - Delimitation and identification of monitored irrigated rice fields.

Time profiles of NDVI (ROUSE, et aL, 1974) were generated during the development cycle of irrigated rice with the aim of mapping the spatial distribution and assessing the uniformity of crops in different periods of development. 10 images from the LANDSAT8/OLI satellite were used, in orbit 224/80 and 225/80 between the months of September and March 2016/2017. The uniformity of the crop in each period analysed was assessed using the coefficient of variation of the NDVI images. Field data validated the information.

RESULTS AND DISCUSSION

Of the 10 images processed, those with the least cloud interference and corresponding to different stages of crop development were chosen for analysis. Figure 2 shows all the images acquired for the period.

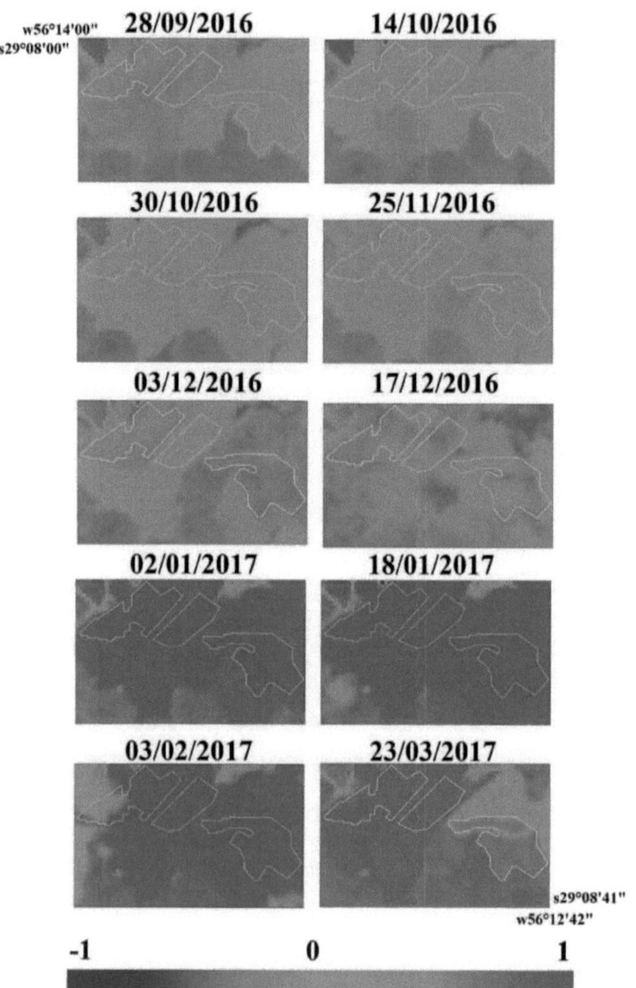

Figura 2   - Multi-temporal NDVI imaging of irrigated rice crops for the 2016/2017 harvest. We selected the images from 25/11/2016 corresponding to the emergence of the cultivars; 03/12/2016 which marks the period of vegetative development; 02/01/2017 and 18/01/2017 corresponding to the reproductive phase of the irrigated rice crop; and the image from 23/03/2017 marked the senescence of the crop. It can be seen that there are no clouds on these dates (Figure 2).

Using the descriptive statistics of the set of NDVI values from the pixels of each image analysed in the different phases of the rice cycle for each crop (Table 1), it can be seen that the spatial variability of the NDVI showed coefficient of variation (CV) values between 4.1 and 5.55% in the emergence period for the two cultivars. In the vegetative development period, the CV increased in both fields with the IRGA 424 RI cultivar, reaching 7.7% and decreasing to 2.89% in the IRGA 429 cultivar.

For the reproductive stage, there is a decrease in variability and a linear characteristic for the two cultivars in the three fields, with values ranging from 1.08% to 2.2% for the two images analysed in this period. This decline is due to the almost maximum coverage of the soil by the canopy and the uniform colouring of the leaves during this period (VERHULST and GOVARETS, 2010).

Senescence was marked by CV values of 3.77% for the IRGA 429 cultivar; 7.6% for the IRGA 424 RI cultivar in the 100 hectare field; and for the 60 hectare field the CV increased, reaching 13.3%.

Table 1 - Descriptive statistics of the set of NDVI values of the pixels in each image analysed at the different stages of the rice cycle for each crop.

Phase/image	Average	Median	Minimum	Maximum	DP	CV(%)	Kurtosis	Ass.
24 HECTARES - IRGA 429								
Emergency (25/11/16)	0.24	0.24	0.21	0.26	0.01	4.10	2.58	-0.28
Vegetative (03/12/16)	0.30	0.30	0.28	0.32	0.01	2.89	2.31	0.26
Reproductive (02/01/17)	0.80	0.80	0.73	0.83	0.02	1.98	4.11	-0.77
Reproductive (18/01/17)	0.84	0.84	0.81	0.86	0.01	1.26	3.04	-0.52
Senescence (23/03/17)	0.67	0.67	0.59	0.71	0.03	3.77	2.55	-0.37
60 HECTARES - IRGA 424RI								
Emergency (25/11/16)	0.33	0.33	0.30	0.38	0.02	5.55	2.49	0.42
Vegetative (03/12/16)	0.46	0.46	0.33	0.55	0.04	7.69	3.46	-0.31
Reproductive (02/01/17)	0.81	0.81	0.76	0.85	0.01	1.78	3.43	-0.63
Reproductive (18/01/17)	0.84	0.84	0.80	0.86	0.01	1.08	4.26	-0.86
Senescence (23/03/17)	0.50	0.51	0.26	0.60	0.07	13.30	4.00	-1.21
100 HECTARES - IRGA 424RI								
Emergency (25/11/16)	0.27	0.27	0.25	0.34	0.01	4.72	6.34	1.13
Vegetative (03/12/16)	0.40	0.39	0.35	0.50	0.02	5.89	5.97	1.29
Reproductive	0.80	0.79	0.75	0.84	0.02	2.20	3.12	-0.12

(02/01/17)							
Reproductive (18/01/17)	0.84	0.84	0.80	0.86	0.01	1.17	3.18	0.00
Senescence (23/03/17)	0.39	0.38	0.31	0.48	0.03	7.60	2.96	0.46

SD = Standard deviation; CV = Coefficient of variation; Ass. = Asymmetry.

During the crop cycle, greater spatial variability was observed in the two fields with the IRGA 424RI cultivar, especially on the dates of 03/12/16 corresponding to the vegetative period and 23/03/17 during senescence. The vegetative period defines the number of panicles per square metre, which can affect rice grain yield (SOSBAI, 2016).

The date 03/12/16 marks the vegetative period 8 days after the cultivar's emergence (25/11/16). This is when tillering begins, when the first nitrogen fertiliser is applied and the water comes in. Irrigation, in the vast majority of crops, is poorly planned, although the water is controlled. Flooding occurs from higher levels, with the water being conveyed by gravity, maintaining a sheet of water by means of ditches built with a difference in level of 5 to 10 cm (SOSBAI, 2016). The higher coefficient of variation of 7% in the vegetative period in the 60-hectare crop with the IRGA 424RI cultivar may be associated with the low flow rate with which water irrigated this crop in this period.

The irrigation water for the 60-hectare field comes from a dam 1.5 kilometres away, while the 100-hectare field is located next to the dam (Figure 1). It is known that the water requirement of rice irrigated by soil flooding is high, but it varies according to climatic conditions, soil attributes and crop management. Losses in canals, the location of the water collection source and the depth of the water table also influence the volume of water required by the crop (SOSBAI, 2016).

According to information from the field, the water flow into the 60-hectare field was low, due to the long distance travelled by the irrigation channels, which may have caused the greatest disuniformity in this period, as shown by the CV (7.69%). This fact may also have influenced the high degree of unevenness at the end of this crop's development cycle, with a CV of over 13%, since the vegetative period analysed can compromise grain yields if there are any anomalies in the crop. These variations identified in these periods (vegetative and senescence) may also be related to the yields of these crops, which were 10474kg/ha in the 100-hectare crop and 9814 kg/ha in the 60-hectare crop.Water management in flood-irrigated rice is fundamental to the performance and uniformity of the crop. As well as influencing the physical appearance of rice plants, water interferes with the availability of nutrients, the population and species of weeds and the incidence of certain insect pests and diseases (SOSBAI, 2016).

CONCLUSIONS

In general, there was greater spatial variability in the three crops analysed at the start of the development cycle, from emergence to the vegetative stage. After this initial phase, the canopy began to close and the VC declined, due to the soil being covered by the plant canopy and the uniformity of leaf colour during this period. At the end of the cycle, the CV increased again.

There is evidence that CVs can serve as uniformity indicators for irrigated rice crops and can be related to the final performance of the crop, especially when the CV is above 7% at the beginning and 13% at the end of the development cycle, which can indicate signs of vegetation stress and possible economic damage.

More studies are needed in different irrigated rice fields to verify the correlation between CV and crop uniformity and, consequently, crop yield.

BIBLIOGRAPHICAL REFERENCES

KUENZER, C.; KNAUER, K. Remote sensing of rice crop areas. **International Journal of Remote Sensing.** 2013. 2101-2139.

ROUSE, J. W. et al. **Monitoring Vegetation Systems in the Great Plains with ERTS.** Third Earth Resources Technology Sattellite-1 Symposium. Greenbelt: NASA. 1974.

SOSBAI, S.B. **Irrigated rice:** technical research recommendations for southern Brazil. ISBN 978-85-69582-02-1. Ed Pelotas, 2016. 200p.

TORBICK, N. et al. Monitoring Rice Agriculture across Myanmar Using Time Series Sentinel-1 Assisted by Landsat-8 and PALSAR-2. **Remote Sensing,** v.9, n. 119, 2017.

VERHULST, N.; GOVARETS, B. The normalised difference vegetation index (NDVI) Green Seeker TM handheld sensor: Toward the integrated evaluation of crop management. Part A: Concepts and case studies. **International Maize and Wheat Improvement Centre- CIMMYT,** Mexico, 2010.

CHAPTER 5

SPECTRAL BEHAVIOUR OF DIFFERENT IRRIGATED RICE CULTIVARS DURING THEIR DEVELOPMENT CYCLE

Cassiane Jrayj de Melo Victoria Bariani,

Nelson Mario Victoria Bariani.

INTRODUCTION

Rice is one of the most important crops globally, with more than half of the world's population depending on rice production for their food supply. It is considered the cereal with the greatest potential for increasing production (SOSBAI, 2016). It is estimated that over the next 40 years, food production will increase by 60 per cent in developed countries and up to 100 per cent in developing countries (ALEXANDRATOS and BRUINSMA, 2012). In Brazil, for example, the state of Rio Grande do Sul (RS) is responsible for 70 per cent of the total produced, which currently guarantees the supply of this cereal to the Brazilian population (SOSBAI, 2016).

The most widely used cultivars in RS, mainly due to their adaptability to the soil and climate and market preference, are IRGA 424RI, Puitá Inta CL, Guri Inta CL and IRGA 409, among others. The development cycle of cultivars in Rio Grande do Sul can vary between super-early, with a cycle of less than 100 days, early, from 110 to 120 days, medium, from 121 to 130 days, and semi-early, greater than 130 days. There are four types of plant architecture: traditional (tall plants), intermediate, semi-dwarf/filipin (modern/filipin) and semi-dwarf/American (modern/American). Distinguishing between plant groups helps growers as it facilitates decision-making regarding the management practices to be adopted, the diagnosis of biotic and abiotic stresses, and whether or not plants are susceptible to lodging (MAGALHÃES and FAGUNDES, 2015).

Remote sensing is useful as a tool to support the monitoring, management or inspection of crops. It uses electromagnetic information reflected by the earth's surface, such as vegetation indices, to identify stress, productivity and/or monitor the growth of crops and agricultural cultivars (GHOBADIFAR, WAYAYOR, *et al,* 2016; BASTIAANSSEN and STEDUTO, 2017; HOMMA, MAKI and HIROOKA, 2017). As a result, methodologies for monitoring cultivars and their development have incorporated this type of data into their modelling and/or field correlations (WANG, HUANG, et a/., 2015).

Methodologies that use satellite remote sensing can help and facilitate the various agents involved in production, as well as various aspects such as crop management, monitoring and inspection.

For the producer, the spatio-temporal distribution of information at the level of agricultural plots allows for remote monitoring of the behaviour of crops throughout their development cycle, resulting in lower travel costs when monitoring commercial crops, as it allows for better planning and use of field trips.

For the industry, obtaining spectro-temporal curves for each cultivar facilitates its identification and traceability, which generates greater precision and accuracy when evaluating the grain, since each cultivar has an added value.For research institutes such as IRGA, for example, both space-time distribution and obtaining spectro-temporal curves help to monitor and validate crops with the aim of seed certification. The aim of this research is therefore to characterise the spectro-temporal curves of the cultivars Puitá Inta CL, Guri Inta CL, IRGA 409 and IRGA424RI in the municipality of Itaqui, RS using remote sensing techniques.

MATERIAL AND METHODS

The study area corresponds to 20 agricultural plots with 4 different cultivars, Puitá Inta CL, Guri Inta CL, IRGA 409 and IRGA424RI, located in the municipality of Itaqui, on the western border of Rio Grande do Sul (Figure 1).

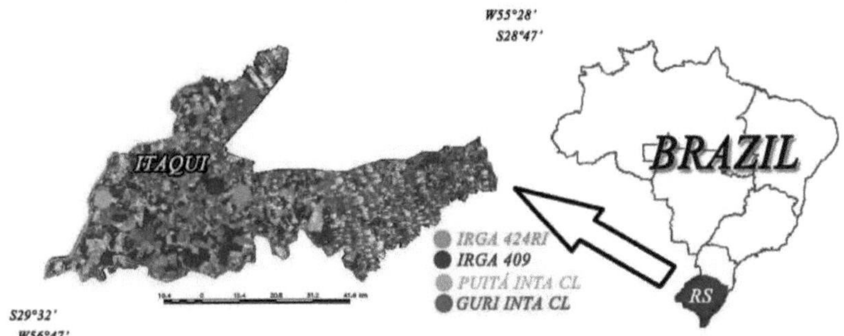

Figure 1- Location map of the study area, Brazil; Rio Grande do Sul; Itaqui; Crops with different cultivars.

To extract the spectral curves of the different cultivars throughout the development cycle, 20 scenes corresponding to orbits 224 and 225 at point 80 of the Landsat8/OLI satellite were used during the period corresponding to the rice cycle (September to March), in the 2016/2017 harvest, in the municipality of Itaqui, RS.

Of the 20 scenes analysed, 10 had a quality standard considered sufficient for spectral analysis, i.e. with low cloud interference.

The images of the study area were subjected to radiometric and geometric corrections. The procedures were carried out in the

Spring GIS from INPE (Câmara, Souza, Freitas, & Garrido, 1996). Firstly, the agricultural plots were registered, contrasted, RGB colour composited and vectorised. Secondly, the reflectance was calculated and the vegetation index extracted. Atmospheric correction was carried out locally on the image, using a procedure adapted from Chavez et al. (1996).There are numerous vegetation indices extracted from satellite images. One of the most widely used and well-known indices is NDVI, the normalised difference vegetation index (ROUSE, HAAS, *et al.,* 1974). Normalisation is carried out using the equation:

$$NDVI = \frac{\rho_{nir} - \rho_{red}}{\rho_{nir} + \rho_{red}} \qquad (1)$$

Where ρ_{nir} is the reflectance value in the near infrared range; and ρ_{red} is the reflectance value in the visible red range.

In order to record the differences in the spectral behaviour of each cultivar, the rate of change was calculated, i.e. the slope, calculated as the difference in NDVI (ZHANG, FRIEDL, *et aL,* 2003). The nomenclature used for each section of the curves was adapted from (ALLEN and PEREIRA, 2009) and (ALLEN, PEREIRA, et a/., 1998), illustrated in figure 2.

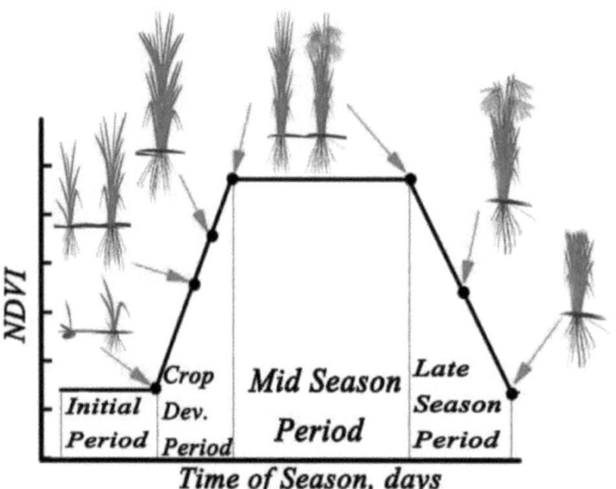

Figure 2 - Nomenclature for each section of the NDVI curves.

Source: Adapted from Allen and Pereira (2009) and FAO 56 (ALLEN, PEREIRA, *etal.,* 1998).

RESULTS AND DISCUSSION

The spectral curves with the NDVI time series for the four rice cultivars are shown in Figure 3. The four rice cultivars showed bell-like curves, as also evidenced in the work of (WANG, ZHANG, *et al.,* 2014) who analysed the spectral characteristics of NDVI throughout the cycle of wheat and rice crops in China.

Each development period can be identified and analysed by considering the increase or decrease in NDVI between consecutive images during the rice crop development period e.g. (BOSCHETTI, STROPPPIANA and BRIVIO, 2009; XIN, YU and VAN LEEUWEN, 2002; WANG, HUANG, *etal.,* 2015).

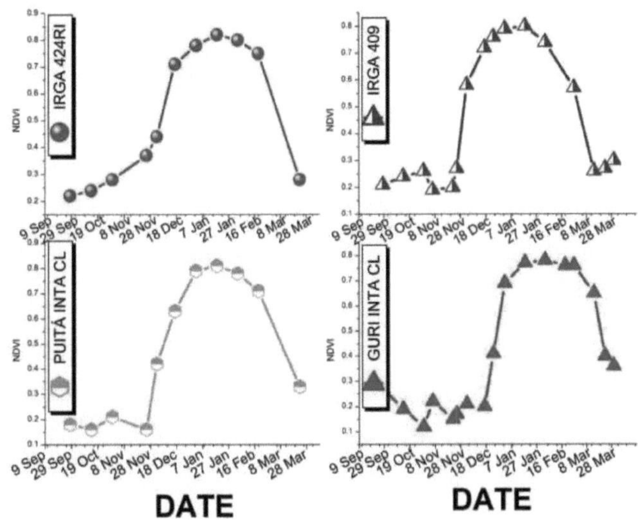

Figure 3- Average NDVI behaviour corresponding to the cultivars IRGA 424RI, IRGA 409, Puitá Inta CL and Guri Inta CL in 20 plots analysed in the 2016/2017 harvest.

NDVI values ranged from 0.22 to 0.82 for the IRGA 424RI cultivar, 0.16 to 0.81 for the Puitá Inta CL cultivar, 0.19 to 0.80 for the IRGA 409 cultivar and 0.12 to 0.78 for the Guri Inta CL cultivar.

In general, the initial period is marked by NDVI values between 0.12 and 0.28. This period was longer for the Guri Inta CL cultivar because it was sown later, at the end of November, while the other cultivars were sown in October and early November.

During the development period, there was a rapid increase in NDVI values, which ranged from 0.37 to 0.71 for the IRGA 424RI cultivar, 0.42 to 0.63 for the Puitá Inta CL cultivar, 0.58 to 0.76 for the IRGA 409 cultivar and 0.41 to 0.69 for the Guri Inta CL cultivar. This rapid growth rate is due to the increase in biomass caused by tillering, stimulated by the first nitrogen fertilisation and the entry of water into the crop (SOSBAI, 2016), which makes nutrients available more quickly due to solubilisation, leading to accelerated vegetative development and consequently an increase in NDVI (WANG, *etaL,* 2015).

The maximum NDVI values for the mid-season period ranged from 0.78 to 0.82 for the IRGA 424RI cultivar, 0.79 to 0.81 for the Puitá Inta CL cultivar, 0.79 to 0.80 for the IRGA 409 cultivar and 0.77 to 0.78 for the Guri Inta CL cultivar. Works such as those by Wang et al., (2014) and Wang et al., (2015) which analysed the spectral behaviour of NDVI in rice fields in Nanjing and Zhenjiang in China also identified the maximum values of vegetation indices occurring during the period of maximum crop development.

During the Late Season, NDVI values decrease to between 0.75 and 0.28 for the IRGA 424RI cultivar, 0.71 and 0.33 for the Puitá Inta CL cultivar, 0.57 and 0.27 for the IRGA 409 cultivar and 0.65 and 0.36 for the Guri Inta CL cultivar. This observed decrease in NDVI values is due to crop senescence or death (WANG, HUANG, *et al.,* 2015) and indicates the approach of harvest (SOSBAI, 2016).

In order to analyse the rate of change of the NDVI spectrum-time curves throughout the rice development cycle for each cultivar, the first derivative was applied. Figure 4 shows the first derivative of the NDVI corresponding to the cultivars IRGA 424RI, IRGA 409, Puitá Inta CL and Guri Inta CL in the 20 plots analysed for the 2016/2017 harvest.

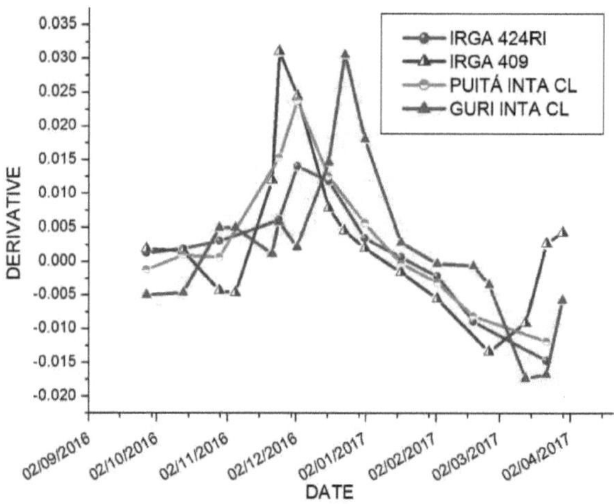

Figure 4 - First derivative of the NDVI corresponding to the cultivars IRGA 424RI, IRGA 409, Puitá Inta CL and Guri Inta CL in 20 plots analysed in the 2016/2017 harvest.

In general, the derivative varied from -0.005 to 0.03 during the development cycle of irrigated rice. During the initial period the slope remains practically zero, around -0.005 to 0.005. The lowest NDVI values are found during this period, which corresponds to soil preparation, sowing and water intake (SOSBAI, 2016). This information corroborates the work of Wang et aL, (2014) and Wang et aL, (2015) who found the lowest values of vegetation indices and zero slope rates in this period which corresponds to soil preparation, water intake and transplanting seedlings.In the development period, the NDVI began to increase rapidly and its slope became greater than 0.005 for the first time. This corresponds to the period of rapid development when biomass increases rapidly due to tillering triggered by the fertilisation and irrigation that was carried out at the end of the previous period (SOSBAI, 2016) and whose results are manifested in this period. Subsequently, the slope reached its maximum value, in the "Mid Season Period", which meant that the NDVI reached its maximum rate and the plant reached its maximum development. Then the NDVI rate began to decrease, in the "Late Season Period", which marks the senescence of the crop, returning to values below zero.

CONCLUSIONS

There is evidence that remote sensing data can help monitor the cultivars IRGA424RI, IRGA409, Puitá Inta CL and Guri Inta CL in commercial plantations in the municipality of Itaqui in the state of Rio Grande do Sul. There was a clear correspondence between the development periods of the different cultivars and the dynamics of the NDVI values.

The first derivative was analysed in the NDVI profile and the rate of change of the NDVI was expressed as the difference between its values in the curve, i.e. its slope. The minima and maxima of these curves were applied to analyse the development periods of the cultivars IRGA 424RI, IRGA 409, Puitá Inta CL and Guri Inta CL. All four development periods described by FAO 56, Initial Period, Crop Dev. Period, Mid Season Period and Late Season Period, could be determined using this method for different rice cultivars, making it possible to support growth and irrigation models.

BIBLIOGRAPHICAL REFERENCES

ALEXANDRATOS, N.; BRUINSMA, J. **World Agriculture Towards 2030/2050: The 2012 revision ESA E Working Paper No. 12-03. FAO.** ROME, ITALY, p. 1-154. 2012. (http://www.fao.org/docrep/016/ap106e/ap106e.pdf).

ALLEN, R. G. et al. FAO Irrigation and drainage paper No. 56. In: FAO **Food and Agriculture Organisation of the United Nations.** Rome: [s.n.], 1998. p. 26-40.

ALLEN, R. G.; PEREIRA, L. S. Estimating crop coefficients from fraction of ground cover and height. **Irrig. Sei.,** 28, 2009. 17-34.

BASTIAANSSEN, W. G. M.; STEDUTO, P. The water productivity score (WPS) at global and regional levei: Methodology andfirst results from remote sensing measurements of wheat, rice and maize. **Science of the Total Environment,** v. 575, 2017. 595-611.

BOSCHETTI, M.; STROPPPIANA, D.; BRIVIO, P. A. Multi-year monitoring of rice crop phenology through time series analysis of MODIS images. **International Journal of Remote Sensing,** v.18, 2009. 4643-4662.

CÂMARA, G. et al. Intergrating remote sensing and GIS by object-oriented data modelling. **Computers & Graphics,** v. 20, n. 3, p. 395-403, May 1996.

CHAVEZ, J. Image-based atmospheric corrections - revisited and improved. **Protogrammetric Engineering and Remote Sensing,** 62, n. 9, 1996. 1025- 1036.

GHOBADIFAR, F. et al. Detection of BPH (brown planthopper) sheath blight in rice farming using multispectral remote sensing. **Geomatocs, Natural Hazards and Risk,** v. 7, 2016. 237-247.

HOMMA, K.; MAKI, M.; HIROOKA, Y. Development of rice simulation model for remote-sensing (SIMRIW-RS). **Journal of Agricultural Meteorology,** v. 73, 2017. 9-15.

MAGALHÃES, A. M. D.; FAGUNDES, P. R. Embrapa. **Embrapa Technological Information Agency,** 2015. Available at: <http://www.agencia.cnptia.embrapa.br/gestor/arroz/arvore/CONTOOOfojvoko c02wyiv80bhgp5povqqj3b.html:*. Accessed on: 06 June 2017.

ROUSE, J. W. et al. **Monitoring Vegetation Systems in the Great Plains with ERTS.** Third Earth Resources Technology Satellite-1 Symposium. Greenbelt: NASA. 1974.

SOSBAI, R. T. D. C. D. A. I. **Irrigated rice:** technical research recommendations for southern Brazil. ISBN 978-85-69582-02-1. ed. Pelotas: [s.n.], 2016. 200 p.

WANG, J. et al. Estimation of rice phenology date using integrated HJ-1 CCD and Landsat-8 OLI vegetation indices time-series images. **Journal of Zhejiang University-SCIENCE B,** v. 16, 14 October2015. p. 832-844.

WANG, L. et al. Multi-Temporal Detection of Rice Phenological Stages Using Canopy Spectrum. **Rice Science,** China, 2014. 108-115.

XIN, J.; YU, Z.; VAN LEEUWEN, L. Mapping crop key phenological stages in the north China plain using NOAA time series images. **International Journal of Applied Earth Observation and Geoinformation,** v. 2, 2002. 109-117.

CHAPTER 6

REMOTE SENSING TECHNIQUES FOR MONITORING THE MAIN PHENOLOGICAL STAGES OF RICE

Cassiane Jrayj de Melo Victoria Bariani,

Nelson Mario Victoria Bariani.

INTRODUCTION

The importance of rice both commercially and nutritionally has led to an increase in its production. It is the staple food of more than three billion people and the second most cultivated cereal in the world, second only to maize. Currently, rice is the crop with the greatest potential for increased production and accounts for 20 per cent of the calories consumed in the world's diet (SOSBAI, 2016).

Brazil ranks $9°$ in the world ranking of countries with the highest rice production and $1°$ among the Mercosur countries (CONAB, 2017). Rio Grande do Sul stands out as the country's largest producer, accounting for 70% of the total produced in Brazil, thus guaranteeing the supply of this cereal to the Brazilian population (SOSBAI, 2016).

Rice in Rio Grande do Sul is produced in 131 municipalities located in the southern half of the state, but it is on the western border that the highest yields are achieved, with the municipalities of Itaqui, Maçambará, São Borja and Uruguaiana standing out with yields that exceed 8,000 kg per hectare (IRGA, 2015). Improvements in the synchronisation between the detection of phenological stages and management procedures could take it to 10 t/ha.

Traditional methodologies for assessing the phenological stages of rice are based on morphological criteria whose application in the production process has limitations in terms of practicality, cost and precision. Remote sensing techniques make it possible to correlate phenological stages with variables whose determination does not present the limitations mentioned above.

Remote sensing is a tool with a high potential to help monitor, manage or inspect crops. Remote monitoring with satellite images is scientifically recognised for providing information that can be associated with the vigour, density, health, nutrition and development of vegetation by conveniently using the electromagnetic information reflected by the earth's surface, using, for example, so-called vegetation indices (TASUMI & ALLEN, 2007; SAKAMOTO, et aL, 2010). As a result, methodologies for estimating development stages have incorporated this type of data into their modelling and/or field correlations (WANG, et aL, 2015). The aim of this research is therefore to estimate the main phenological stages of the irrigated rice crop using remote sensing techniques associated with modelling to monitor the development cycle, with a view to the main managements in commercial plantations in the municipality of Itaqui in the state of Rio Grande do Sul.

MATERIAL AND METHODS

In this study, time profiles of NDVI (ROUSE, et aL, 1974) were generated during the development cycle of the IRGA 424RI cultivar, which has a medium cycle (131-135 days), with the aim of mapping the distribution of plant phenology following the scale described by Counce et al. (2000) in three commercial irrigated rice fields in the municipality of Itaqui, located on the western border of the state of Rio Grande do Sul.

We used 10 images from the LANDSAT8/OLI satellite in orbits 224/80 and 225/80 between September and March 2016/2017. The NDVI values were evaluated based on the temporal profiles and associated with the phenological stages obtained using the GD Rice model (STEINMETZ et aL, 2015). Field data obtained from producers, technical assistance or IRGA, depending on availability, was used to validate the forecasts.

The GD Rice model has proved to be efficient for estimating phenological stages and supporting the management of rice crops in Rio Grande do Sul. It can be accessed online for free at http://www.cpact.embrapa.br/agromet (STEINMETZ, et aL, 2015). The model uses data from official meteorological stations of the National Institute of Meteorology (INMET), and the São Borja meteorological station was used in this work.

RESULTS AND DISCUSSION

The occurrence of phenological stages is highly variable as it depends on temperature, the plant's nutritional status and the type of cultivar (STANSEL, 1975). For this reason, it is preferable to express the stages of plant development in days, but estimated using degree days (GD), or thermal sum (STRECK, et aL, 2006), and then associate them with calendar days. In this research, NDVI was associated with phenological stages, taking GD into account using the GD Rice model with the support of information on sowing and plant emergence in three fields planted with the IRGA 424RL cultivar.

The main development stages assessed by remote sensing were: sowing, emergence, vegetative stage V4, reproductive stage R1, R2, R4, R8 and R9. Figure 1 shows the graph of NDVI values throughout the development cycle of the IRGA 424RI cultivar, with the dates of each image analysed and their respective phenological stages illustrated.

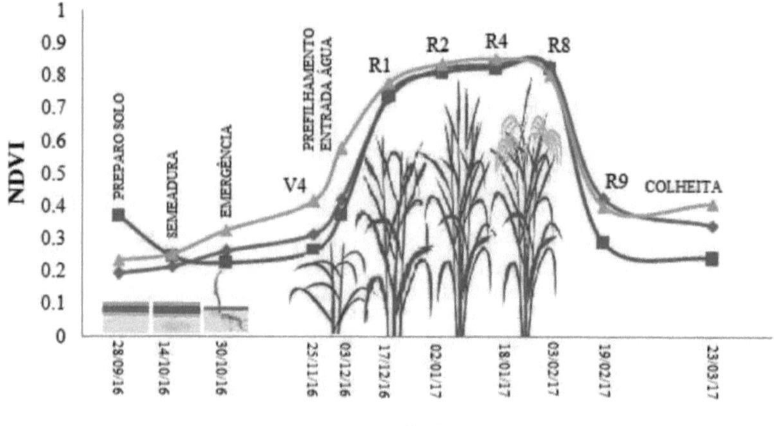

Figure 1 - Correspondence between NDVI and phenological stages for the IRGA 424RI cultivar in three commercial crops analysed in the 2016/2017 harvest.

The NDVI from sowing to rice seedling emergence showed the lowest values, ranging from 0.19 to 0.3. Vegetation index values lower than 0.3 in this period may be associated with a reduction in vegetation following the application of herbicides used to control weeds (NOBRE, 2010).

Vegetative stage V4 occurred on 25 November 2016 when the plants had four leaves and the NDVI values were above 0.3 (0.31 to 0.41). One week after the occurrence of stage V4, on 3 December 2016, the plants began to tiller. From the V4 stage onwards, the first nitrogen fertiliser topdressing and irrigation water are applied (SOSBAI, 2016), so the accelerated start of vegetative growth is expected.

On 17 December 2016, the crops were in reproductive stage R1 with NDVI values between 0.73 and 0.78. At this stage, the panicle is differentiating and the second nitrogen fertiliser is being top-dressed
(SOSBAI, 2016), so further growth in biomass is expected from the irrigated rice crop.

The next stage, R2, took place on 2 January 2017, two weeks after R1, with NDVI values between 0.81 and 0.84. In R2 the panicle begins to expand within the stem, reaching its maximum. At this stage, the division of the pollen grain mother cells is taking place, and it is one of the most critical in the plant's development to the occurrence of stresses, especially those caused by low temperature (below 17 °C) and nutrient deficiency (SOSBAI, 2016).

Flowering began on 18 January 2017 when the plants were at R4 and the NDVI values peaked at between 0.83 and 0.85. Some authors report that high vegetation indices saturate (HUETE, et aL, 2002), for example, (GU, et aL, 2013) observed that high leaf area index (LAI) values made the NDVI insensitive. In this study, however, this was not the case.

The next phenological stage of great importance for monitoring is R8, when ripening begins. This stage began on 3 February, indicating the approach to harvest and when NDVI values began to decline, falling below 0.83. Full ripeness (R9) occurred a week after the start of ripeness, on 11 February 2017, and from this point on it was possible to harvest. The image that would have represented stage R9 takes place a week after its occurrence on 19 February 2017, when the NDVI was between 0.29 and 0.42.

As for the effect of saturation, it should be noted that it is not a limiting factor in identifying phenological stages for the IRGA 424RI cultivar, as the highest NDVI values occur between R4 and R7 and decrease from R8 onwards, a stage that is important to identify due to its proximity to harvest. The robustness and great potential of NDVI for identifying the main phenological stages of rice was also demonstrated by Wang et. aL, (2015). There is therefore evidence that the main phenological stages of the IRGA 424RI cultivar can be identified from the temporal profiles of vegetation indices, as also demonstrated by the work of Nobre (2010) and Wang, et. aL, (2015).

CONCLUSIONS

There is evidence that remote sensing data can help monitor the main phenological stages of irrigated rice for the IRGA 424 RI cultivar in commercial plots in the municipality of Itaqui in the state of Rio Grande do Sul.The scale of the study, commercial agricultural plots, proved to be adequate, as well as not presenting problems of saturation of the NDVI values, since the main phenological stages of the IRGA 424RI cultivar were clearly identified.

BIBLIOGRAPHICAL REFERENCES

CONAB, C. N. D. A. National Supply Company, 2017. Available at: <www.conab.gov.br>. Accessed on: 11 March 2017.

COUNCE, P. A.; KEISLING, T. C.; MITCHELL, A. J. A Uniform, Objective, and Adaptive System for Expressing Rice Development. **Crop Sei.,** v. 40, 2000. 436-443.

GU, Y.; WYLIE, B. K.; HOWARD, D. M. NDVI saturation adjustment: a new approach for improving cropland performance estimates in the Greater Platte River Basin, USA. **Ecol. Indic.,** v. 30, 2013. 1-6.

HUETE, A. R.; DIDAN, K.; MIURA, T.; RODRIGUEZ, E. P.; GAO, X; FERREIRA, G. OverView of the Radiometric and Biophysical Performance of the MODIS Vegetation indices. **Remote Sensing of Environment,** 83, 2002. 195-213.

IRGA. **2014/2015 CROP REPORT.** GOVERNMENT OF THE STATE OF RIO GRANDE DO SUL. ITAQUI, p. 1-3. 2015.

NOBRE, F. L. D. L. **SPECTROTEMPORAL CHARACTERISATION OF IRRIGATED RICE CROPS USING MODIS IMAGES.** PELOTAS: UFPEL, 2010.

SOSBAI, R. T. D. C. D. A. I. **Irrigated rice:** technical research recommendations for southern Brazil. ISBN 978-85-69582-02-1. ed. Pelotas: [s.n.], 2016. 200 p.

STANSEL, J. W. **The rice plant:** its development and yield. In: SIX DECADES OF RICE RESEARCH IN TEXAS. Beaumont: Texas Agricultural Experiment Station. 1975. p. 9-21.

STRECK, N. A.; BOSCO, L. C.; MIICHELON, S.; ROSA, H. T.; WALTER, L. C.; PAULA, G. M.; CAMERA, C.; LAGRO, L; MARCOLIN, E. Evaluation of the response to photoperiod in irrigated rice genotypes. **Bragantia,** Campinas, v. 65, n. ed. 4, 2006. p. 533-541.

WANG, JING; HUANG, JING-FENG; WANG, XIU-ZHEN; JIN, MENG-TING; ZHOU, ZHEN; GUO, QIAO-YING; ZHAO, ZHE-WEN; HUANG, WEI-JIAO; ZHANG, YAO; SONG, XIAO-DONG. Estimation of rice phenology date using integrated HJ-1 CCD and Landsat-8 OLI vegetation indices time-series images. **Journal of Zhejiang University-SCIENCE B,** v. 16,14 October2015. p. 832-844.

CHAPTER 7

MAPPING AND CALIBRATION OF NDVI FOR ESTIMATING THE PHENOLOGY OF CULTIVAR IRGA 424 RI

Cassiane Jrayj de Melo Victoria Bariani,

Nelson Mario Victoria Bariani.

INTRODUCTION

Monitoring crop phenology and growth makes it possible to understand the actual patterns of crop development and to identify critical stages of development and their relationship to various management practices (VINA et al., 2004), including irrigation, nutrition and phytosanitary interventions. In general, these monitoring tasks are costly and time-consuming and often subject to great uncertainty due to the extrapolation of specific observations to large areas (BASTIAANSSEN, 2000). It is therefore valuable to describe the spectral behaviour of crops, cultivars and varieties, analysing typical characteristics of each one, with a view to identifying the main stages of development throughout the crop cycle (SAKAMOTO *et al.,* 2010).

Knowing when the different phenological stages occur in irrigated rice crops is of great importance for planning the management practices to be used. The main phenological stages to be considered in irrigated rice crops are: stage V4; reproductive stage R1 (panicle differentiation); reproductive stage R2 (rubberisation) to R4 (start of flowering); reproductive stage R8 (start of senescence); and reproductive stage R9 (maturation) (SOSBAI, 2016).

In general terms, the development cycle of the IRGA 424 RI cultivar is average, with full flowering at 103 days. These are plants with strong stalks and good architecture, with short, erect and hairy leaves, tolerance to excess iron in the soil, high yield potential and resistance to Brusone in the leaf and panicle. It has low initial vigour, is resistant to lodging, has tillering capacity and can produce more than 10t/ha. In the Western Frontier of Rio Grande do Sul, average yields exceed 8t/ha

(IRGA, 2015). The limiting factors for increasing this productivity and filling the so-called "productivity gap" in relation to maximum potential are believed to be linked to crop management. From this point of view, it is justifiable to make an effort to carry out more accurate monitoring of the main phenological stages of irrigated rice and their spatial distribution within the cultivated plot or field in order to carry out management practices at the appropriate phenological stage according to the physiological and agronomic characteristics of each cultivar (STEINMETZ *et al.*, 2015). The aim is therefore to help map the main phenological stages using vegetation index throughout the cycle of the IRGA 424 RI cultivar in commercial irrigated rice fields.

MATERIAL AND METHODS

In this study, NDVI time profiles (ROUSE *et al.*, 1974) were generated during the development cycle of the IRGA 424RI cultivar in order to map the distribution of plant phenology in commercial irrigated rice fields in the municipality of Itaqui, located on the western border of the state of Rio Grande do Sul.

Eight images from the LANDSAT8/OLI satellite in orbit 224/80 were used between September and March 2016/2017. The NDVI values were evaluated from the temporal profiles and associated with the degree days of development (GD) obtained from the GD Rice model (STEINMETZ *et al.*, 2015). Field data obtained from producers, technical assistance or IRGA, depending on availability, was used to adjust the model and validate its predictions.The GD Rice model has proved efficient for estimating phenological stages and supporting the management of rice crops in RS. It can be accessed online for free at http://www.cpact.embrapa.br/agromet (STEINMETZ *et al.*, 2015). The model uses data from official meteorological stations of the National Institute of Meteorology (INMET) which are not available for some municipalities, such as Itaqui and Maçambará, which is why the nearest meteorological station (São Borja) was used. A correlation between GDD and days after emergence (DAE) was also developed using information from the same sources mentioned above.

RESULTS AND DISCUSSION

The main phenological stages of the IRGA 424 RI cultivar were identified by associating NDVI with GDD and DAE in the commercial irrigated rice fields analysed. Table 1 shows the dates imaged by the Landsat8/0LI satellite, the main development stages of the IRGA 424 RI cultivar, the number of days of the phenological stages and the date of occurrence of each stage estimated by the GD Rice model, together with the correlation with DAE obtained using data from the INMET weather station in São Borja.

Table 1. Dates of images from the Landsat8/0LI satellite, orbit 224 point 80. Estimates by the GD Rice model of the main stages of development of the irrigated rice crop, number of days of duration of each phenological stage (by correlation), as well as their date of occurrence.

DATES	NDVI		PHENOLOGICAL	MODEL GD RICE	
LANDSAT8/0 LI IMAGES	A	B	STAGES	N° DAYS	DATES
28/09/201	0.3	0.2	Soil preparation	0	28/10/20
14/10/201	0.2	0.2	Sowing	0	14/10/20
30/10/201	0.2	0.2	Emergency	10	24/10/20
25/11/201	0.2	0.3	V4	16	09/11/20
03/12/201	0.4	0.4	Tillering/	-	03/12/20
17/12/201	0.4	0.4	R1	58	21/12/20
02/01/201	0.8	0.8	R2	70	02/01/20
18/01/201	0.8	0.8	R4	89	21/01/20
03/02/201	0.8	0.8	R8	110	11/02/20
19/02/201	0.8	0.8	R9	118	19/02/20
23/03/201	0.6	0.5	Harvesting	150	22/03/20

The average NDVI values for each phenological stage in the two fields analysed can be seen in Table 1. In addition, the spatio-temporal distribution of NDVI values can be seen in the maps in Figure 1, which can then be associated with the phenological stages.

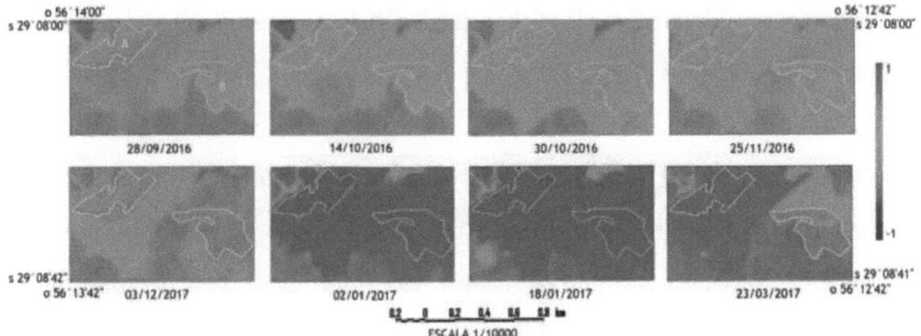

Figure 1 - Spatio-temporal distribution map of NDVI values throughout the irrigated rice crop cycle.

As an example, Figure 1 shows that on 28/09/2016 field A was in the soil preparation phase, while B was already being sown; on 14/10/2016 both fields were already sown; in the image from 30/10/2016 both fields are in the emergence phase; on 25/11/2016 both fields are in the V4 stage; in the image of 03/12/2016 the two fields are tillering, but there is the beginning of reddish colouring in some pixels, especially in field B, indicating that it is more advanced in relation to field A, or some type of stress may be occurring in the vegetation; on 02/01/2017 to 18/01/2017 the fields are in reproductive stages, reaching the maximum NDVI values (0. 84) as shown in Table 1.84) as shown in Table 1. In these stages, the predominance of red colour indicates that the plants are uniform; finally, in the image from 23/03/2017, the fields are already mature and harvesting is starting, where the NDVI values decrease, and it is possible to see the start of harvesting in field B.

The extraction of the average NDVI values corresponding to the plots (Table 1) added to the visualisation of the distribution of pixels within each field analysed (Figure 1) made it possible to assess the development of the vegetation and possible anomalies occurring within the fields due to the non-uniformity detected mainly in the images from 03/12/2016 and 23/03/2017.

To validate the uniformity, the coefficient of variation of the NDVI values between all the pixels within each field was extracted. In the image of 03/12/2016, which corresponds to tillering and the entry of water into the crop, the fields showed a coefficient of variation of 5.9% for field A and 7.7% for field B; and for the image of 23/03/2017, the coefficients of variation were 3.5% and 13.3% in the respective fields.

There is evidence that field B was less uniform compared to field A and that this may be associated with management, due to the date of sowing, as shown in the image of 28/09/2016, as well as the existence of probable water, nutritional or management deficiencies in the V4 phenological stage, tillering and the entry of water into the crop observed in the image of 03/12/2016; and finally at full maturity and the start of the harvest on 23/03/2017, also evidenced by the coefficient of variation of 13.3%.

CONCLUSION

The NDVI values associated with the degree days of development made it possible to differentiate between the main phenological stages of the IRGA 424 RI cultivar throughout the irrigated rice crop cycle in Itaqui.

Mapping and visualising the spatial distribution of NDVI values proved to be a powerful tool for identifying weak points in the crop, serving as a basis for future planning and management of the crops analysed.

BIBLIOGRAPHICAL REFERENCES

IRGA. **2014/2015 CROP REPORT.** GOVERNMENT OF THE STATE OF RIO GRANDE DO SUL. ITAQUI, p. 1-3. 2015.

JENSEN, J. R. **Remote sensing of the environment:** a perspective on terrestrial resources. [S.l.]: Parenthesis, v. 2, 2011. 598 p. Authorised translation.

LOPES, S. I. G. et al. **CULTIVAR IRGA 429:** ANOTHER OPTION FOR THE PRE-GROWN RICE CULTIVATION SYSTEM IN RIO GRANDE DO SUL. BRAZILIAN CONGRESS ON IRRIGATED RICE. PELOTAS: [s.n.]. 2015. p. 1-5.

ROUSE, J. W. et al. **Monitoring Vegetation Systems in the Great Plains with ERTS.** Third Earth Resources Technology Satellite-1 Symposium. Greenbelt: NASA. 1974.

SAKAMOTO, T. et al. A Two-Step Filtering approach for detecting maize and soybean phenology with time-series MODIS data. **Remote Sensing of**

Environment, 114, 2010. 2146-2159.

SOSBAI, R. T. D. C. D. A. I. **Irrigated rice:** technical research recommendations for southern Brazil. ISBN 978-85-69582-02-1. ed. Pelotas: [s.n.], 2016. 200 p.

STEINMETZ, S. et al. **GD Rice: A Degree-Day Based Programme to Support Planning and Decision-Making in Irrigated Rice Management.** Embrapa Temperate Climate. Technical Circular, 162. Pelotas, p. 1-8. 2015 (ISSN 1516-8832).

STEINMETZ, S. et al. **GD Rice: A Degree-Day Based Programme to Support Planning and Decision-Making in Irrigated Rice Management.** Embrapa. Pelotas, p. 1-8. 2015 (ISSN 1516-8832).

CHAPTER 8

MAPPING THE MAIN PHENOLOGICAL STAGES OF THE IRGA 409 CULTIVAR USING REMOTE SENSING TECHNIQUES

Cassiane Jrayj de Melo Victoria Bariani,

Nelson Mario Victoria Bariani.

INTRODUCTION

The development cycle of the IRGA 409 cultivar is average, lasting between 121 and 130 days from emergence to maturity, with panicle primordium at 65 days, full bloom at 89 days and physiological maturity at 126 days. They are low-growing plants with short, erect, hairy leaves, a panicle protected by the flag leaf, susceptible to toxicity from excess iron in the soil and sensitive to low temperatures. It has high initial vigour, is resistant to lodging, has a high tillering capacity and can produce more than 8t/ha (SOSBAI, 2016).

Identifying the main phenological stages throughout the crop cycle can be used to better plan the main management tasks to be carried out on the crop. Methodologies that use remote sensing to map and monitor crop cycles have been increasingly used over the last three decades in academic work, but are still little explored in agricultural management in Brazil. The NDVI *(Normalised Difference Vegetation Index)* is an index obtained using remote sensing techniques based on the contrast between the maximum absorption in the red band by the chlorophyll present in the plant and the maximum reflectance in the near infrared due to the leaf's cellular structure (JENSEN, 2011), serving as an indicator of crop growth and health. The NDVI value, properly filtered of interference, describes the spectral behaviour of crops and can be associated with the characteristics of cultivars and varieties. The differentiation produced by the typical characteristics of each plant canopy during growth potentially makes it possible to identify stages of development throughout the agricultural crop cycle (SAKAMOTO *etal.,* 2010).

Knowledge of the time of occurrence of the different phenological stages in irrigated rice farming is of great importance for planning the management practices to be used. The main phenological stages to be considered in irrigated rice crops are: stage V4 (collar formed on the fourth leaf of the main stalk); reproductive stage R1 (panicle differentiation); reproductive stage R2 (rubberisation) to R4 (beginning of flowering); reproductive stage R8 (beginning of senescence); and reproductive stage R9 (maturation) (SOSBAI, 2016).In the Western Frontier of Rio Grande do Sul, the average yield exceeds 8t/ha (IRGA, 2015). It is believed that if improvements are made to crop management, this yield could be increased. To this end, it is advisable to monitor the main phenological stages of irrigated rice more accurately and their spatial distribution within the cultivated plot or field in order to carry out management practices at the appropriate phenological stage according to the physiological and agronomic characteristics of each cultivar (STEINMETZ *et al.*, 2015). The aim is therefore to help map the main phenological stages using vegetation index throughout the cycle of the IRGA 409 cultivar in commercial irrigated rice fields.

MATERIAL AND METHODS

Surface reflectance data was used from 11 images from the LANDSAT8/OLI satellite, orbit 224 at point 80, between September and March 2016/2017, obtained from https://earthexplorer.usgs.gov . The NDVI was calculated according to Rouse *et al.* (1974), and graphed as a function of time to evaluate the temporal profiles and associated with the degree days of development (GDD). The phenological stages were associated according to the degree days or corresponding days after emergence, obtained from the GD Rice model (STEINMETZ *et al.*, 2015) and empirical correlations between days and degree days for the 2016-2017 harvest obtained from regional INMET data.

The NDVI time profiles of the IRGA 409 cultivar from two commercial irrigated rice fields in the municipality of Itaqui, RS, were analysed during the development cycle in order to map the distribution of plant phenology. The phenology and productivity data came from the producer's technicians or from technical assistance or from IRGA, depending on availability. The GD Arroz model was accessed via the internet at: http://www.cpact.embrapa.br/agromet (STEINMETZ *etal.*, 2015). The phenological stages and the degree days required for their manifestation were obtained from this model and then associated with the NDVI values corresponding to the same GDD.

RESULTS AND DISCUSSION

The main phenological stages of the IRGA 409 cultivar were identified by associating NDVI with GD in the four commercial irrigated rice fields. Table 1 shows the dates of imaging by the Landsat8/OLI satellite, the main stages of development of the IRGA 409 cultivar, the number of days of duration of the phenological stages and the date of occurrence of each stage estimated by the GD Rice model using the INMET weather station in São Borja.

Table 1. Dates of the Landsat8/OLI satellite images, orbit 224 point 80. Estimates by the GD Rice model of the main stages of development of the irrigated rice crop, number of days of duration of each phenological stage, as well as their date of occurrence.

DATES LANDSAT8/OLI IMAGES	NDVI		PHENOLOGICAL STAGES	MODEL GD RICE	
	A	B		N° DAYS	DATES
28/09/2016	0.2	0.2	Soil preparation	0	28/10/2
14/10/2016	0.2	0.2	Sowing	0	10/10/2
30/10/2016	0.2	0.2	Emergency	10	20/10/2
25/11/2016	0.2	0.3	V4	17	05/11/2
03/12/2016	0.5	0.6	Tillering/	-	03/12/2
17/12/2016	0.7	0.7	R1	58	16/12/2
18/01/2017	0.7	0.7	R4	89	17/01/2
03/02/2017	0.7	0.6	R8	109	06/02/2
19/02/2017	0.5	0.6	R9	119	16/02/2
26/02/2017	0.5	0.3	Senescence	129	
14/03/2017	0.2	0.2	Senescence	135	
23/03/2017	0.2	0.3	Harvesting	From	22/03/2

The ranges of average NDVI values for the plots analysed, for each phenological stage, in the two fields analysed, according to the degree days obtained by the GD Rice model, converted to days after emergence (DAE) by correlation between GDD and DAE for the 2016-2017 season, are shown in Table 1.

The spatio-temporal distribution of NDVI values and, correspondingly, the phenological stages of the IRGA 409 cultivar can be seen in the maps in Figure 1 over the course of the crop year in question.

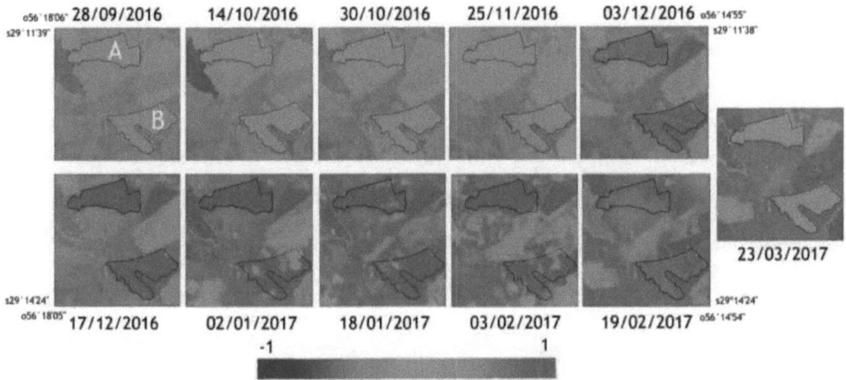

Figure 1 - Spatiotemporal distribution map of NDVI values throughout the irrigated rice crop cycle, cultivar IRGA 409, associated with the phenological stages in Table 1.

Figure 1 shows that on 28/09/2016 fields A and B were in the soil preparation stage; on 14/10/2016 both fields were already sown; in the image from 30/10/2016 both fields are in the emergence stage; on 25/11/2016 both fields are in the V4 stage; in the image from 03/12/2016 both fields are tillering and the start of reddish colouring can be seen; from 17/12/2016 to 18/01/2017 both fields are in the reproductive stage, reaching the maximum NDVI values (0.76) as shown in Table 1. In these stages the predominance of red colour indicates that the plants are uniform; on the dates between 03/02/2017 and 19/02/2017 the NDVI begins to decline indicating the complete maturation of the crop which indicates the proximity to harvest; finally in the image of 23/03/2017 the fields were already harvested which is observed through the greenish colour corresponding to NDVI values between 0.27-0.34.

By extracting the average NDVI values (Table 1) and visualising the distribution of pixels within each field analysed (Figure 1), it was possible to assess the development of the vegetation. In general, a uniformity was observed within the two fields analysed, this fact also being evidenced by the NDVI coefficient of variation, which showed a value of 10.7% for field A and 13.7% for field B. This uniformity revealed by the statistical results was reflected in the yield, which reached 8718.6 kg/ha in field A and 8202.36kg/h in field B. Although validating the calibration between NDVI and phenological stages with field data is beyond the scope of this study, we can say that the existing data coincided well with the data inferred by the model.

CONCLUSION

The NDVI values associated with the degree days of development made it possible to differentiate between the main phenological stages of the IRGA 409 cultivar throughout the irrigated rice crop cycle in Itaqui. The mapping and visualisation of the spatial distribution of NDVI values proved to be a powerful tool and can serve as a basis for future planning and management of the irrigated rice crops analysed.

BIBLIOGRAPHICAL REFERENCES

IRGA. **2014/2015 CROP REPORT.** GOVERNMENT OF THE STATE OF RIO GRANDE DO SUL. ITAQUI, p. 1-3. 2015.

JENSEN, J. R. **Remote sensing of the environment:** a perspective on terrestrial resources. [S.l.]: Parêntese, v. 2, 2011. 598 p. Authorised translation.

LOPES, S. I. G. et al. **CULTIVAR IRGA 429:** ANOTHER OPTION FOR THE PRE-GROWN RICE CULTIVATION SYSTEM IN RIO GRANDE DO SUL. BRAZILIAN CONGRESS ON IRRIGATED RICE. PELOTAS: [s.n.]. 2015. p. 1-5.

ROUSE, J. W. et al. **Monitoring Vegetation Systems in the Great Plains with ERTS.** Third Earth Resources Technology Satellite-1 Symposium. Greenbelt: NASA. 1974.

SAKAMOTO, T. et al. A Two-Step Filtering approach for detecting maize and soybean phenology with time-series MODIS data. **Remote Sensing of Environment,** 114, 2010. 2146-2159.

SOSBAI, R. T. D. C. D. A. I. **Irrigated rice:** technical research recommendations for southern Brazil. ISBN 978-85-69582-02-1. ed. Pelotas: [s.n.], 2016. 200 p.

STEINMETZ, S. et al. **GD Rice: A Degree-Day Based Programme to Support Planning and Decision-Making in Irrigated Rice Management.** Embrapa

Temperate Climate. Technical Circular, 162. Pelotas, p. 1-8. 2015 (ISSN 1516-8832).

STEINMETZ, S. et al. **GD Rice: A Degree-Day Based Programme to Support Planning and Decision-Making in Irrigated Rice Management.** Embrapa. Pelotas, p. 1-8. 2015 (ISSN 1516-8832).

CHAPTER 9

SPECTRO-TEMPORAL IDENTIFICATION OF IRRIGATED RICE CULTIVARS USING SATELLITE IMAGES

Cassiane Jrayj de Melo Victoria Bariani,
Nelson Mario Victoria Bariani.

INTRODUCTION

Monitoring spectral characteristics over time using satellite images is scientifically recognised for providing information that can be associated with the type of cultivar or variety, vigour, density, health, nutrition and phenological stages, as it conveniently uses the electromagnetic information reflected by the earth's surface, using, for example, so-called vegetation indices.

Knowing the spectral signature of different irrigated rice cultivars is an important tool for monitoring and inspecting crops, as it can facilitate the work of public certification bodies and rice industries. The Riograndense Rice Institute (IRGA) is the organisation that monitors and inspects irrigated rice plantations in southern Brazil, guaranteeing the multiplication of certified seeds, verifying the location of the plantation, the size of the cultivated area, the cultivar used and whether red rice is present.

Methodologies that make use of remote sensing can utilise the spectral records contained in the different images (bands) obtained by satellite sensors, and process them to transform them into products capable of contributing to aspects such as crop management, monitoring and inspection, and facilitating technically-based decision-making. For the producer, the spatiotemporal distribution of information at the level of agricultural plots can enable remote monitoring of crop behaviour throughout their development cycle, resulting in lower travel costs when monitoring commercial crops, as it enables better planning of field trips. For the industry, the analysis of spectro-temporal curves for each cultivar makes it possible to characterise them in advance, which generates greater precision and accuracy when receiving and evaluating the rice, since each cultivar and even each crop can be considered to have a different added value. For research institutes such as IRGA, for example, both space-time distribution and obtaining spectro-temporal curves contribute to monitoring and validating crops for the purpose of seed certification.

The aim of this study was to monitor and describe the spectrotemporal curves of the IRGA 409, IRGA 424, IRGA 426, IRGA 429 and IRGA 430 cultivars planted in commercial crops in the municipality of Itaqui, by combining data obtained in the field and temporal NDVI profiles throughout the crop cycle, obtained by processing 11 images from the LANDSAT8/OLI satellite, in orbits 224/80 and 225/80 corresponding to the municipality of Itaqui, RS.

The spectral curves throughout the cycle showed particular spectral behaviour between the different cultivars, assessed using curve parameters obtained at reference points at the start of growth, the rapid growth stage, the stage of maximum NDVI, and the senescence stage.

MATERIAL AND METHODS

The study area corresponds to 81 agricultural plots with 5 different cultivars, IRGA 409, IRGA 424, IRGA 426, IRGA 429 and IRGA 430 (Figure 1).

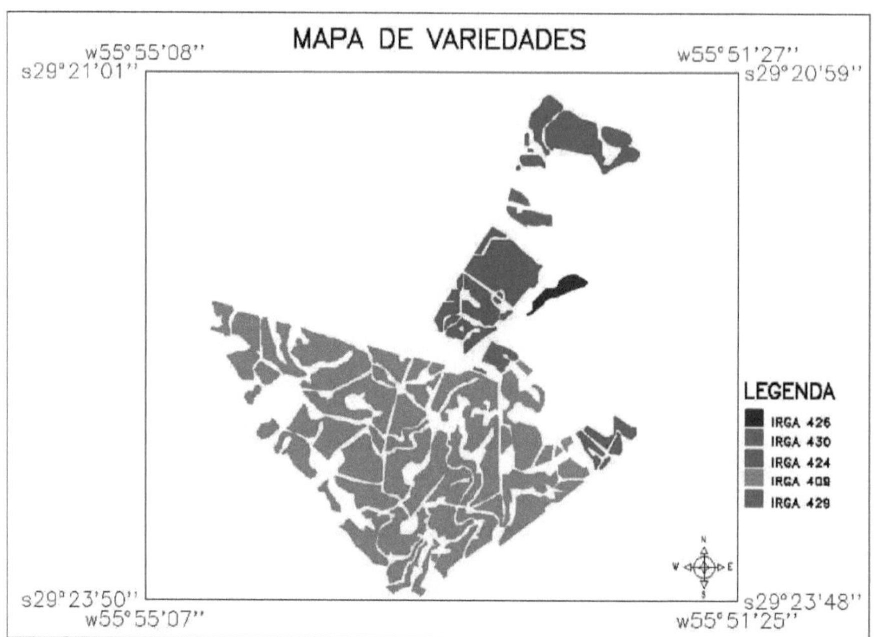

Figure 1 - Map of the different varieties for the 103 agricultural plots analysed.

To extract the spectral curves of the different cultivars throughout the development cycle, 11 scenes were obtained corresponding to orbits 224 and 225 at point 80 of the Landsat8/OLI satellite during the period corresponding to the rice cycle (September to March), in the 2016/2017 harvest, in the municipality of Itaqui, RS.

The images of the area under study were subjected to radiometric and geometric corrections and NDVI extraction (ROUSE *et aL,* 1974). The procedures were carried out using Spring software. Firstly, procedures were carried out to register, contrast and obtain RGB images as a basis for vectorising the agricultural plots, as well as calculating reflectance, atmospheric correction and extracting the vegetation index. Secondly, the average NDVI for each plot was extracted for each image and NDVI curves were constructed over time.The sections of each curve for each cultivar were analysed following the methodology described in Hargrove *et aL,* (2010), where the points with the minimum and maximum NDVI on each side of the graph are found and the points with NDVI equal to the minimum NDVI increased by 20% and 80% of the difference ($NDVI_{Max}$- $NDVI_{min}$) of the rapid development and senescence stages are calculated, as well as the range of points with NDVI max. Both the NDVI values and the time periods in days between the different reference points mentioned were recorded in tables.

RESULTS AND DISCUSSION

Using the NDVI extracted from the Landsat8/OLI images over the course of the irrigated rice development cycle, it was possible to identify the characteristics of the curves by means of reference points whose NDVI values and the duration of each stage in days are shown in Table 1.

Table 1. Reference point values for analysing the different development periods during the cycle of the different cultivars.

	NDVI					DAYS				
	409	424	426	429	430	409	424	426	429	430
ESQ 20%	0.39	0.38	0.39	0.37	0.36	45	45	59	58	46
ESQ 80%	0.69	0.71	0.70	0.72	0.72	29	34	13	18	31
MAX	0.79	0.80	0.81	0.83	0.83	23	18	24	20	22
MAX	0.80	0.82	0.80	0.84	0.84	15	15	5	16	13
DIR 80%	0.71	0.73	0.74	0.78	0.75	20	19	28	17	17
DIR 20%	0.47	0.49	0.57	0.58	0.51	29	30	12	38	11

The initial 20% period, from sowing to the start of growth, lasted 45 and 46 days for the cultivars IRGA 409, 424, 430; and 58 and 59 days for the cultivars IRGA 429, 426. Lower NDVI values of between 0.36 and 0.39 were observed for all cultivars, due initially to the soil exposed at sowing and the application of herbicides (NOBRE, 2010) and later the entry of water. During this period, the first nitrogen application and water intake must have already taken place (SOSBAI, 2016). The water layer causes NDVI values to vary, while nitrogen fertilisation makes nitrogen available to the seedlings (more quickly due to solubilisation), leading to accelerated vegetative development marked by the start of rapid growth and consequently an increase in NDVI (WANG, *etaL,* 2015).

At the 80 per cent or rapid growth stage, the increase in biomass was already quite significant, which led to an increase in NDVI values, which ranged from 0.69 to 0.72. This period occurred between two and four weeks after the start of growth (minimum 20%), as shown in Table 1. At this stage, there is total coverage of the soil/water blade by the vegetation (ALLEN and PEREIRA, 2009), and it is quite likely that the crop is entering the reproductive stage (R1), which is characterised by the differentiation of the floral primordium. At this stage, the second nitrogen fertilisation takes place and the number of spikelets in each panicle is defined (SOSBAI, 2016).

The maximum NDVI values occurred during the crop's ripening period (HARGROVE, *et al.*, 2010), when vegetative growth ceases and full flowering of the crop begins, at which point most of the plants (main stalks and tillers) have their panicles exposed and their spikelets open (SOSBAI, 2016). NDVI values in this period ranged from 0.79 to 0.84 (Table 1). This moment is characterised by the greatest risk of yield loss, and the ambient temperature cannot reach 17°C or less (NOBRE, 2010). The duration of the ripening stage varied between 29 and 38 days depending on the cultivar (Table 1).

After the NDVI values reach their maximum, they decline, which is characterised by the senescence or death of the crop (WANG, *et al.*, 2015). The beginning of this period indicates that the harvest is approaching, and all that is needed is the ideal humidity, close to 22%, for the grain to be harvested (SOSBAI, 2016). NDVI values fell to between 0.71 and 0.78.

At harvest, the NDVI values return to lower levels, as shown in the graph by the area on the right (minimum 20%) with NDVI values between 0.47 and 0.58 (Table 1).

From the above, there is evidence that the shape of the curve as well as the NDVI values and the duration of the different development periods can be a support tool for identifying cultivars, varieties and phenological stages of the rice crop planted in each plot or field analysed using Landsat images.

CONCLUSION

There is evidence that the temporal NDVI profile of irrigated rice crops can support the identification of different cultivars and can be used by the industry, seed certification bodies and/or funding agencies as a tool to support their mandatory monitoring and inspection.

In order to use images from sensors on board satellites for agricultural monitoring purposes, it is necessary to obtain as many cloud-free images as possible during the harvest, which can be difficult. In this study, the fact that the study area was in an overlapping area of two Landsat8 satellite orbits made it easier to take scenes, totalling twenty images, making it possible to discard nine that were interfered with by clouds and still have eleven favourable images for monitoring the harvest. Therefore, Landsat8/OLI satellite images when in overlapping orbit areas become a suitable tool for agricultural monitoring.

BIBLIOGRAPHICAL REFERENCES

ALLEN, R. G.; PEREIRA, L. S. Estimating crop coefficients from fraction of ground cover and height. **Irrig. Sei.,** 28, 2009. 17-34.

HARGROVE, W. W. et al. Toward a National Early Warning System for Forest Disturbances Using Remotely Sensed Land Surface Phenology. **USGS,** 2010. Available at: <https://www.geobabble.org/~hnw/first/ncdc/slideshow.html>. Accessed on: 13 June 2017.

NOBRE, F. L. D. L. **Temporal Spectral Characterisation of Irrigated Rice Crops using Modis Images.** PELOTAS: UFPEL, 2010.

ROUSE, J. W. et al. **Monitoring Vegetation Systems in the Great Plains with ERTS.** Third Earth Resources Technology Satellite-1 Symposium. Greenbelt: NASA. 1974.

SOSBAI, R. T. D. C. D. A. I. **Irrigated rice:** technical research recommendations for southern Brazil. ISBN 978-85-69582-02-1. ed. Pelotas: [s.n.], 2016. 200 p.

WANG, J. et al. Estimation of rice phenology date using integrated HJ-1 CCD and Landsat-8 OLI vegetation indices time-series images. **Journal of Zhejiang University-SCIENCE B,** v. 16, 14 October2015. p. 832-844.

I want morebooks!

Buy your books fast and straightforward online - at one of world's fastest growing online book stores! Environmentally sound due to Print-on-Demand technologies.

Buy your books online at
www.morebooks.shop

Kaufen Sie Ihre Bücher schnell und unkompliziert online – auf einer der am schnellsten wachsenden Buchhandelsplattformen weltweit! Dank Print-On-Demand umwelt- und ressourcenschonend produzi ert.

Bücher schneller online kaufen
www.morebooks.shop

info@omniscriptum.com
www.omniscriptum.com

Printed by Books on Demand GmbH, Norderstedt / Germany